# eTELEMED 2022

The Fourteenth International Conference on eHealth, Telemedicine, and Social Medicine

June 26th – 30th, 2022

Porto, Portugal

**eTELEMED 2022 Editors**


Yoshitoshi Murata, Iwate Prefectural University, Japan

Arian Rajh, Agency for Medicinal Products and Medical Devices of the Republic of Croatia, Croatia

Bobby Gheorghiu, Canada Health Infoway, Canada

**Printed from e-media with permission by:**

Curran Associates, Inc.
57 Morehouse Lane
Red Hook, NY  12571



Printed with permission by Curran Associates, Inc. (2023)

International Academy, Research, and Industry Association (IARIA)
412 Derby Way
Wilmington, DE 19810

Phone: (408) 893-6407
Fax: (408) 527-6351

petre@iaria.org

**Additional copies of this publication are available from:**

Curran Associates, Inc.
57 Morehouse Lane
Red Hook, NY 12571 USA
Phone: 845-758-0400
Fax: 845-758-2633
Email: curran@proceedings.com
Web: www.proceedings.com

**Table of Contents**

# eTELEMED 2022

# Forward

The Fourteenth International Conference on eHealth, Telemedicine, and Social Medicine (eTELEMED 2022), held in Porto, Portugal, June 26 - 30, 2022, considered advances in techniques, services, and applications dedicated to a global approach of eHealth.

Development of wireless homecare, of special types of communications with patient data, of videoconferencing and telepresence, and the progress in image processing and date protection increased the eHealth applications and services, and extended Internet-based patient coverage areas. Social and economic aspects as well as the integration of classical systems with the telemedicine systems are still challenging issues.

eTELEMED 2022 provided a forum where researchers were able to present recent research results and new research problems and directions related to them. The topics covered aspects from classical medicine and eHealth integration, systems and communication, devices, and applications.

We take this opportunity to thank all the members of the eTELEMED 2022 Technical Program Committee as well as the numerous reviewers. The creation of such a broad and high-quality conference program would not have been possible without their involvement. We also kindly thank all the authors who dedicated much of their time and efforts to contribute to the eTELEMED 2022. We truly believe that, thanks to all these efforts, the final conference program consists of top quality contributions.

This event could also not have been a reality without the support of many individuals, organizations, and sponsors. We are grateful to the members of the eTELEMED 2022 organizing committee for their help in handling the logistics and for their work to make this professional meeting a success.

We hope that eTELEMED 2022 was a successful international forum for the exchange of ideas and results between academia and industry and for the promotion of progress in eHealth and Telemedicine research. We also hope that Porto provided a pleasant environment during the conference and everyone saved some time for exploring this beautiful city

**eTELEMED 2022 Chairs**

**eTELEMED Steering Committee**
Yoshitoshi Murata, Iwate Prefectural University, Japan
Arian Rajh, Agency for Medicinal Products and Medical Devices of the Republic of Croatia, Croatia

**eTELEMED 2022 Publicity Chair**
Laura Garcia, Universitat Politècnica de València (UPV), Spain
José Miguel Jiménez, Universitat Politecnica de Valencia, Spain

# eTELEMED 2022

# COMMITTEE

**eTELEMED Steering Committee**
Yoshitoshi Murata, Iwate Prefectural University, Japan
Arian Rajh, Agency for Medicinal Products and Medical Devices of the Republic of Croatia, Croatia

**eTELEMED 2022 Publicity Chair**

Laura Garcia, Universitat Politècnica de València (UPV), Spain
José Miguel Jiménez, Universitat Politecnica de Valencia, Spain

**eTELEMED 2022 Technical Program Committee**

Don Adjeroh, West Virginia University, USA
Giovanni Albani, Istituto Auxologico Italiano - IRCCS, Italy
Basel Almourad, College of Technological Innovation, Dubai, UAE
Domingos Alves, Ribeirao Preto Medical School | University of Sao Paulo (USP), Brazil
Prima Oky Dicky Ardiansyah, Iwate Prefectural University, Japan
Theodoros N. Arvanitis, Institute of Digital Healthcare | University of Warwick, UK
Tunç Aşuroğlu, Başkent University, Ankara, Turkey
Rafael Avila, Hospital Universitario Privado de Cordoba, Argentina
Mansoor Baig, King Faisal Specialist Hospital & Research Center, Riyadh, Saudi Arabia / ICIMTH, Greece
Panagiotis D. Bamidis, School of Medicine - Aristotle University of Thessaloniki, Greece
Oresti Baños, University of Granada, Spain
Ivana Bartoletti, Gemserv, UK
Christian-Alexander Behrendt, GermanVasc Research Group/ University Medical Center Hamburg-Eppendorf, Germany
Hrvoje Belani, Ministry of Health - Directorate for e-Health, Zagreb, Croatia
Elisabetta Benevento, University of Pisa, Italy
Arriel Benis, HIT- Holon Institute of Technology, Israel
Sid-Ahmed Berrani, Ecole Nationale Polytechnique, Algiers, Algeria
Vilmos Bilicki, University of Szeged, Hungary
Lucia Billeci, National Research Council of Italy | Institute of Clinical Physiology, Pisa, Italy
Eman Buhagiar, Middlesex University, Malta
Marco Buzzelli, University of Milano - Bicocca, Italy
Enrico Gianluca Caiani, Politecnico di Milano, Italy
Manuel Campos Martínez, University of Murcia, Spain
Nicola Carbonaro, University of Pisa, Italy
Ayan Chatterjee, The University of Agder, Grimstad, Norway
Darwyn Chern, Copa Health, Phoenix, USA
Bhargava Chinni, University of Rochester, USA
Mario Ciampi, National Research Council of Italy | Institute for High Performance Computing and Newtorking, Italy
James J. Cimino, Informatics Institute - University of Alabama at Birmingham, USA
Javier Civit, Cober SL, Spain / Gnomon Informatics, Greece

**Copyright Information**

# Implementation of IoT Data Analysis in Personal Elderly Home Care System

Chuan-Zong Huang
Department of Institute of Mechatronic
Engineering of NTUT
Taipei, Taiwan
Email: t109408005@ntut.edu.tw

Jung-Tang Huang
Department of Institute of Mechatronic
Engineering of NTUT
Taipei, Taiwan
Email: jthuang@ntut.edu.tw

*Abstract*—**At present, the proportion of the elderly population in the world is gradually increasing, but the fertility rate is very low. In the future, the elderly population will be much larger than the young population. More care institutions and systems are needed to share the pressure of the young population. This is a major issue in the future. However, many studies are now on the development of wearable devices and the implementation of non-care Internet of Things (IoT). The subjects must wear a number of uncomfortable sensors and bulky equipment, which is very difficult for the subjects. unfriendly, or developed care-related IoT systems without effectively utilizing the data collected. Therefore, this system combines a complete and novel Internet of Things architecture to effectively collect relevant physiological information of users through Bluetooth bracelets, Bluetooth security charms, depth cameras, and Google Home Nest Mini, and integrate these reliable and diverse Data processing and storage, detailed analysis of every tiny and available information in the data, effective analysis of the subjects' life patterns, and establishment of a personalized care model.**

***Keywords-IoT;Smart Healthcare;Big Data;Wearable;Cloud computing;Data Analysis***

## I. INTRODUCTION

The purpose of big data analysis is to analyze large and complex data sources and discover trends, patterns, customer behaviors and market preferences. With various sensors in the Internet of Things and the Internet that plays an important role, an Internet of Things architecture is formed, resulting in a large number of compound data [2]. Through statistical calculation and data analysis, important information was previously ignored because only a piece of data can be obtained, and from a large amount of composite data and then through big data analysis, deep learning to obtain unknown related information. This research will focus on data analysis of smart home and elderly care applications, focusing on providing smart home and elderly living and improving services, using the data collected by sensors and Google Cloud Services applications to conduct a series of verification, processing, classification, storage, calculation, statistics and analysis are presented in a visual interface. The analysis results of these data can help to observe and adjust the user's related activities and formulate plans on dietary health, sleep quality, life health management and so on.

The main purpose of the care IoT application is to provide intelligent services. This proposal combines powerful cloud services to process a large amount of data to make the care environment more comfortable, convenient, and secure. It also uses data processing and analysis to achieve a personal system. This is the ultimate goal of a smart home for IoT applications [1]. Most of the care and medical IoT systems use existing data for analysis and machine learning [12], and some use simple circuit boards for data collection and focus on cloud computing and algorithm optimization [10] [11], or collect data through too many sensors and instruments, resulting in the burden of action. In this experiment, the combination of light and simple Bluetooth devices and cloud applications, and data analysis tools are used to complete the analysis of physiological information, diet and sleep.

Today's cloud services are divided into the following three categories: Software as a Service (SaaS), Platform as a Service (PaaS) and Infrastructure as a Service (IaaS). This research is based on PaaS, establishes a service-oriented cloud computing architecture, and provides users with SaaS so that users can use our existing services through simple settings, no need to consider storage, management, etc., carried out by the cloud unified management.

The purpose of this research is to focus on the data collection and analysis of the cloud-based care system of the Internet of Things. We used a combination of physiological bracelets, wearable amulet, Google Home Nest Mini and other various sensors to form a complete care Internet of Things (IoT) system. A complete and detailed collection of various physiological information, posture behavior and eating habits data were collected through the Bluetooth sensor and store it in the cloud database. We also stored the collected physiological data information, posture changes and design life dialogue questions and answers. Finally, we analyzed the collected physiological information and compared it with the content of the dialogue questionnaire. First, we obtained the difference between the user's subjective psychology, physiological feelings and physiological information. Further, we analyzed the user's living habits, eating habits, sleeping habits, etc., Based on it, we gave active suggestions through audio, and made daily health reports and weekly report analysis, providing doctors to judge the change of user's habits, and maximize the analysis. The approach helped the utilization of all kinds of big data collected, so as to achieve the ultimate goal of personalized care by the application of data in the Internet of Things system. Compared with reference [4], it can be found that through long-term behavior records, it is possible to understand what habits such as bad diet, sleep and so on before getting sick, which may lead to problems and proactively issue reminders and suggestions.

Finally, through a stable system and long-term records, four characteristics that meet the main points of big data are obtained:

- Volume: a lot of data.
- Variety: many different forms of data, unstructured and structured.
- Velocity: the frequency of incoming data.
- Veracity: the credibility of the data.

And analyze it through the following steps:

- Identify analysis goals.
- Data collection.
- Data preprocessing.
- Data analysis.
- Visual representation.

Through the above processing, we can get the highest value analysis results. However, our consideration of data collection may not be comprehensive enough, and users may use the device incorrectly, which will lead to the risk of data abnormality or loss. We need more accurate methods to judge the data. And in the case of collecting a large amount of data, it may cause the equipment to be overloaded or generate redundant data, which will be the goal we need to consider and solve.

Through the above introduction, we clearly understand the market demand and the purpose of the experiment, and we also discussed the initial system architecture and method. At the beginning of the second part, we will start to introduce the complete hardware device and system architecture. The third part will introduce the architecture of the cloud service and the dialogue flow. The fourth part will introduce the experimental procedure, data processing and complete data analysis.

## II. SYSTEM ARCHITECTURE

This section will introduce a flowchart of the IoT system architecture, including devices, cloud services, and dialog flows.

### A. *System Architecture*

The system uses Bluetooth wearable devices, Bluetooth sensors, depth cameras, video pens and other Bluetooth devices, through the Node Network (the Node Network transmission process can perform indoor positioning and tracking) to automatically sense the sensors and push the data to the terminal through the Bluetooth mesh network (Bluetooth Router and IoT Edge Device), then authenticate the Bluetooth device through IoT Edge Device and transmit the data to the cloud database through W-Fi for processing, classification, storage, calculation and statistics, and finally, through other visualizations such as web pages interface for viewing.

The overall system architecture is mainly divided into six levels, and the architecture diagram is abstracted as shown in Figure 1. Each level from top to bottom is:

- Sensor: Responsible for sensing the measured information and converting the sensed information into specific information output. In this system, new Bluetooth devices can be added at any time for data collection
- Node network: A mesh network composed of Bluetooth is responsible for transmitting sensor information to the router through the Bluetooth mesh network.
- Bluetooth router: It is responsible for information transfer and manages the entire sensor network group, which is one of the important roles in the architecture.
- IoT Edge device: It can be said to be a gateway, responsible for controlling data flow, and connecting directly to the cloud to provide filtering, collecting, uploading data and control services.
- Cloud services: Push system services to the world through the public cloud. In this system, new cloud services can be added at any time for more diverse data processing
- User interface: Application(APP) 、 Raspberry Pi, Google Home Nest Mini、Web.

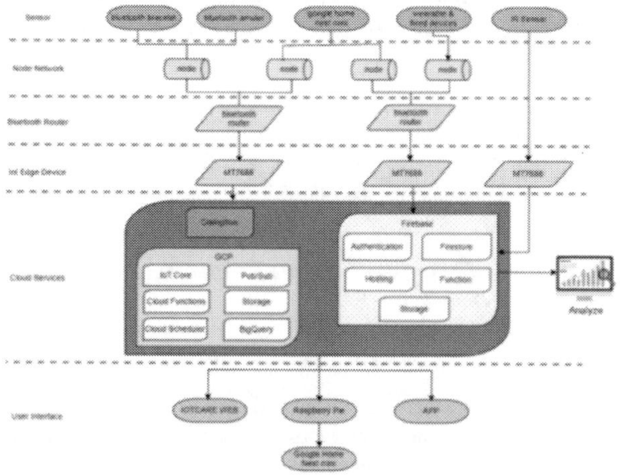


Figure 1. System Architecture Diagram.

### B. *Bluetooth Device and Sensor*

The sensors used in this system are mainly divided into 3 major categories: wearable devices, stationary devices, Google Home Nest Mini.

The wearable devices are Bluetooth bracelets, Bluetooth amulets and Bluetooth tags, as shown in Figure 2. The Bluetooth watch is responsible for detecting physiological data, including heart rate per minute, body temperature, steps and calories. The Bluetooth amulet is responsible for posture monitoring, motion monitoring, height monitoring, and fall monitoring. It can clearly determine and record the user's current body posture changes and make notifications through the speaker and LINE APP in a timely manner. In particular, the Bluetooth Safety Amulet uses the 3D printed appearance and specially designed "gecko feet" are pasted on the user's chest, which can most accurately judge the user's posture changes. The Bluetooth tag is responsible for detecting the indoor mobile location of the user and its positioning.

Fixed sensors are touch sensors, and Google Home Nest Mini. A touch sensor is a weight machine, which is triggered by physical contact. Google Home Nest Mini is triggered by the user's voice message. They are all triggered by the same user. The difference is in how they are triggered.

Google Home Nest Mini runs based on Google Cloud Platform (GCP)[8] services, helping us create a dialogue system to process audio messages, processing flow: Voice input, Dialogflow, Agent, Intent, Fulfillment, Intent, Agent, Dialogflow, Voice output, can respond to different dialogue intentions, events, responses are classified, and their classification data are stored. When an abnormal situation occurs in the physiological data stored in Firebase Firestore [8] (for example, the state changes to a fall or an abnormally high heart rate), the speaker will broadcast a critical message in real-time, and transmit the information to the administrator through LINE.

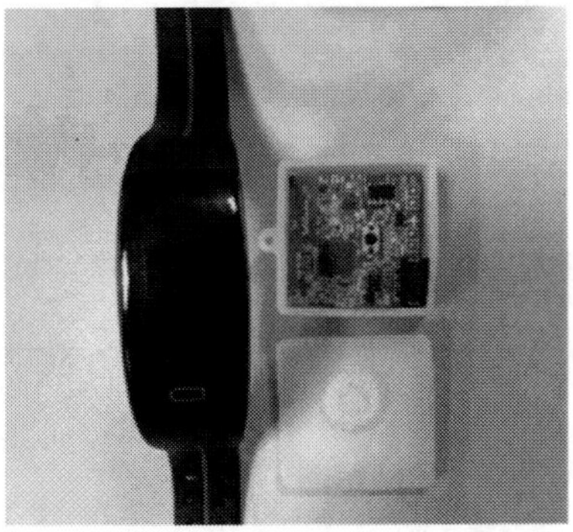

Figure 2. Bluetooth watch and Bluetooth amulet.

### C. Depth Camera

The depth camera used in this system is Intel's RealSense Depth Camera D435. Compared to the Microsoft Kinect series and Leap Motion Controller series, D435 has more detailed long-distance depth detection and has strong inter-SDK compatibility. As shown in Figure 3, this camera is used to take pictures of the user's three meals of food, and use the positioning function of the amulets and the attitude change to trigger the photo. Learn the types of different foods through algorithms, and calculate the volume, calories and nutrients of individual foods, record them in the database, and calculate the daily nutrient intake. For comparison, if the intake is insufficient, dietary advice will be given through the Google Home Nest Mini the next day.

Figure 3. Intel's RealSense Depth Camera D435.

### D. Audio and Video Recording Pen

The audio and video recording pens used in this system are V082S. It has high-quality video and enough memory space, and the battery is enough for a full day of use. The main function is to capture the users eating status outside the main mealtime, and use the RSSI change status to trigger the camera to take pictures. Through the algorithm, the type of each food can be known, and the individual food can be calculated. calorie, and record it in the database, analyze its calorie intake throughout the day, and give appropriate suggestions through the Google Home Nest Mini.

### E. Bluetooth Mesh Network

The Bluetooth LAN is mainly composed of two parts: bNode and bwRouter, which play two important roles in the Bluetooth network, Central (Master) and Peripheral (Slave). The main function of the Master is to scan for the broadcast channel, the address of the Bluetooth tag can be known through the communication protocol defined by this research, and the data is transmitted to the slave through UART, and the slave plays the role of broadcast, responsible for the positioning of the device and the priority of data forwarding and continued transmission. Go to the next bNode until the Master scans its own return packet (Acknowledgement, Ack) and informs the Slave to stop broadcasting. This design can also reduce the delay to ensure whether each data reaches the next bNode. Figure 4 is a relationship diagram between bNodes.

BwRouter is composed of Bluetooth and Raspberry Pi. It is the transmission terminal in the Bluetooth local area network. It acts as a transmission bridge between Bluetooth and Wi-Fi. BLE transmits data to Raspberry Pi through the UART communication interface. Raspberry Pi is filtered, sorted and classified. After that, upload it to the cloud for data storage and analysis through Wi-Fi. BwRouter can achieve a two-way communication transmission system by returning an acknowledgment message (Acknowledgement, Ack) or issuing commands to the lower-level device of the Bluetooth LAN.

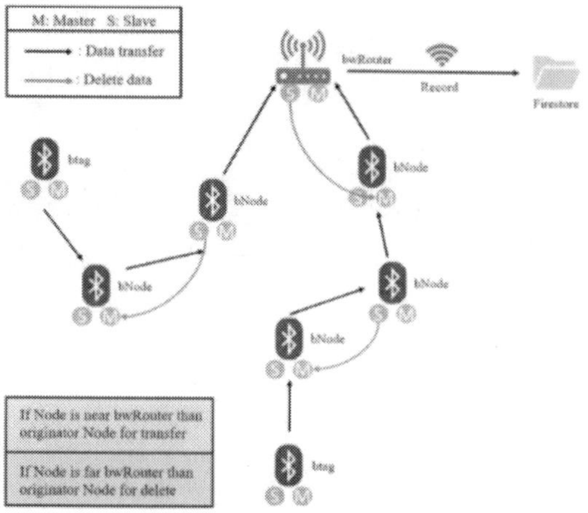


Figure 4. bNode transfer process diagram.

### F. IOT Edge Device

The core of an edge computing device is to act as an entry point or an exit point and to control the data flow at the boundary or periphery between two networks, any hardware responsible for processing data, while edge computing is a large-scale service that was originally handled entirely by the data center. Decompose, cut into smaller and more manageable parts, and distribute to edge nodes for computing [3]. The main reason for using edge computing is that edge computing devices are closer to users than data centers, which can speed up data processing, improve response time, and improve bandwidth availability. It will be closer to the source of the data, so it is more suitable for processing big data and analysis.

### G. BLE Initial Setting

The mobile phone application plays the role of the setting terminal in this system, and the mobile phone application is used to write the relevant information of bNode and bwRoute, so as to make the flooding of the Bluetooth network directional (Flooding is used to publish and relay the message ), and use the built-in latitude and longitude of Google Map to locate the device, so as to know the location of bwRouter, distinguish the area of Mesh network and write the required information, so that Node can judge the distance and have directionality, in addition to setting the latitude and longitude In addition to positioning, field division can also be performed, such as dining room, toilet, living room, bedroom and other areas. As shown in Figure 5.

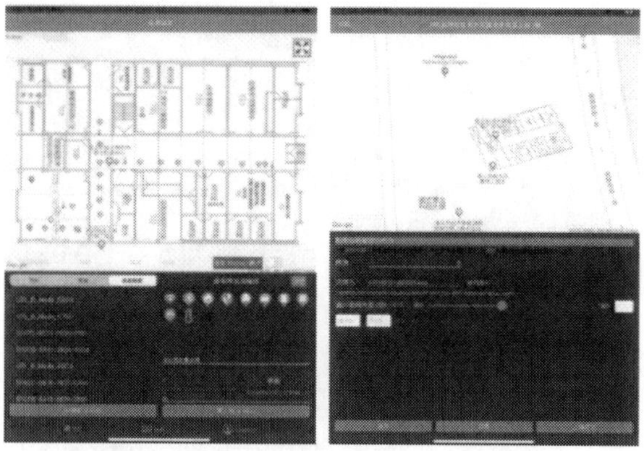

Figure 5. APP interface.

### H. Analyze

Behavior pattern analysis: Physiological data (heartbeat, body temperature, calories, number of steps, posture, movements, etc.) can be obtained through the Bluetooth wearable device to determine behavior patterns, such as daily moving route, which can be used to know whether to go to the toilet, The number and time of filling water, eating or watching TV, and can analyze whether the number of drinking water is insufficient, whether the number of times to go to the toilet is abnormal, etc., and give reminders, and rely on the heartbeat, posture, height changes and location to determine whether it falls or not As well as the location of the fall, it will immediately notify other members of the family through the Google Home Nest Mini, and make appropriate measures immediately.

Sleep quality analysis: You can know the sleep posture change and breathing rate through Bluetooth amulets. Combined with the heartbeat and body temperature changes of the bracelet, you can simply analyze the sleep quality. If there is an abnormal heartbeat change or a large posture change Further in-depth analysis can be carried out. Combined with the question and answer of the Pittsburgh Sleep Quality Scale [6], it can be further confirmed with users, and the results can be analyzed by deep learning training.

Analysis of eating habits: Use the positioning function and posture change of the amulet to trigger the depth camera to shoot the contents of the three meals, and use the audio and video recording pen to capture the eating status outside the mealtime, and conduct in-depth training to identify the types of food and calculate their calorie and nutrient content. The number of weekly exercises, height, weight, age, and gender calculates the individual basal metabolic rate (BMR). The calorie intake per meal obtained by the depth camera is compared with the BMR to determine whether the daily intake of sufficient nutrients is sufficient. And analyze the picky eating behavior, and suggest giving reminders of supplementary nutrients in a timely manner.

Exercise habit analysis: Through the Bluetooth wearable device, you can determine whether you are in a state of exercise. If so, record the duration of the exercise and the total calories consumed after the exercise, and give suggestions for exercise intensity, duration, etc. after each exercise. And record long-term exercise habits. If the habits change or there is no exercise habit, exercise suggestions will be given by analyzing the data.

Based on the above four types of analysis results, it is found that through long-term observation, users can be very familiar with the daily life and various habits of users, even those habits that users themselves have not noticed. Take care of the big and small things in the user's life to achieve the purpose of care and companionship. Finally, the application of deep learning can be added to more deeply simulate the user's habits and status, and establish a personalized digital avatar.

## III. Cloud System Architecture

This research relies on the application of a large number of cloud services and uses part of the cloud functions of Google's cloud platform GCP to classify, organize, store and analyze the data collected by wearable devices, and use the development framework of web pages and related tools. Combined with the personnel positioning and life data collected by the device, it is stored in the object database for web page and APP access, and the data regularly output statistical reports through Extract-Transform-Load (ETL). Then we will analyze the user's behavior, and send it to the front-end community and smart speakers through the push broadcast platform. The cloud architecture of this system is shown in Figure 6.

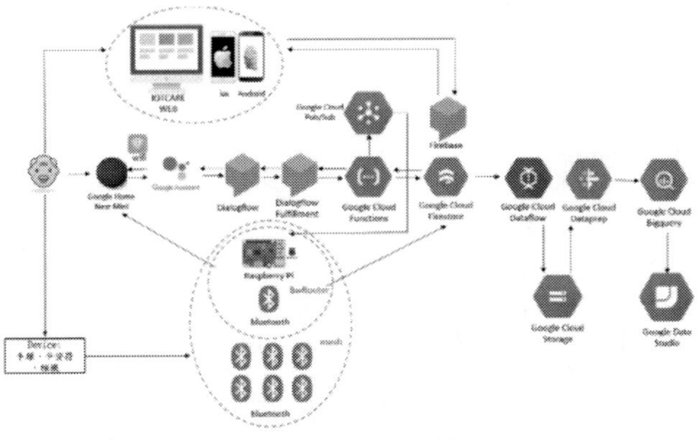

Figure 6. Cloud system diagram.

In this system, we use Google Home Nest Mini to communicate with users, such as proactive reminders, proactive push messages, and in order to collect useful messages through conversations and Q&As with users, we have written many functions and designed different Conversation flow templates. For example, about the Geriatric Depression Scale [6] and the Pittsburgh Sleep Quality Scale [7], we have a preliminary understanding of their emotions and sleep problems, and by analyzing the content of the scales, physiological signals and behavior patterns, we can identify problem points and improve them. factor. The dialogue flow takes the question and answer of the Pittsburgh Sleep Quality Scale [7] as an example, as shown in Figure 7.

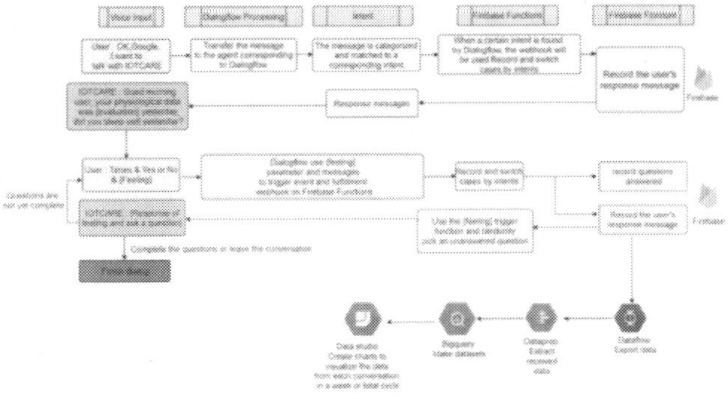

Figure 7. Dialogue questions and answer process.

The cloud services used in this cloud system are described as items E-L in Section 3 of [8], respectively using the cloud service functions shown in Figure 6 to perform data trigger functions, data storage, processing, analysis, visual presentation, etc., and through Google's services, such as Action on Google, Google Home, etc., to deploy and use cloud services.

## IV. Experimental Steps And Data Analysis

This experiment integrates the above-mentioned system architecture and cloud system architecture, and then conducts experiments and data collection, and finally analyzes and validates the data.

### A. *Experimental Field Environment*

The experimental field is a 45 x 31 meters area located on the 7th floor of the Tzung-He hall, which is in the National Taipei University of Technology. The 31 Bluetooth light source devices (bNodes) are subordinated as shown in Figure 8. A one-week (7-day) experiment was conducted from 12:00 pm to 5:00 pm every day to collect each subject's physiological data, posture changes, and daily lunch status, and to analyze each person's physiological information, behavior patterns, diet records and stored in firebase Firestore. Finally, place the Google Home Nest Mini in the center of rooms 709-2 and 709-3 for notification of various conditions.

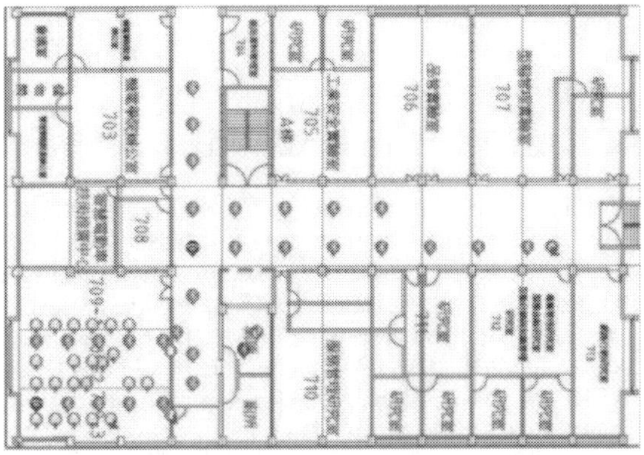

Figure 8. Experimental field and bNode location.

### B. *Data and Data Model*

As shown in Figure 9 the following are noun description and explanations:

- Personal account: Record user permissions and personal related information, user information covers individuals and organizations.
- Notify & Alert: Record alert notifications due to abnormal user status or behavior.
- People: When the user is an individual, this node records all the physiological data information of the current user. Its child nodes include: bracelets, amulets, diet records, exercise records, respectively record physiological and behavioral posture information, and continue to add diet and exercise habits information.
- Location: Record one or more addresses. Its child nodes include floor records, inheriting the address information of the previous floor, and combining multiple floor information, including the latitude and longitude of the floor boundary and the floor plan of the floor.
- Tags: This is a wearable device. The current wearable device of the system is the bracelet and safety charm, which records the current address of the wearable device, floors, mobile fixed equipment. Its child nodes include tags record, which records the historical information of the previous wearable device, including positioning latitude and longitude information.
- Equipment: Inherit the information of the previous layer and record the real-time status of the fixed equipment of this layer. Fixed device information includes weight scales, reed switches, etc., as well as mobile devices, such as tablets, etc., and its child nodes include Device information, the content of which is the contact history information of the device.

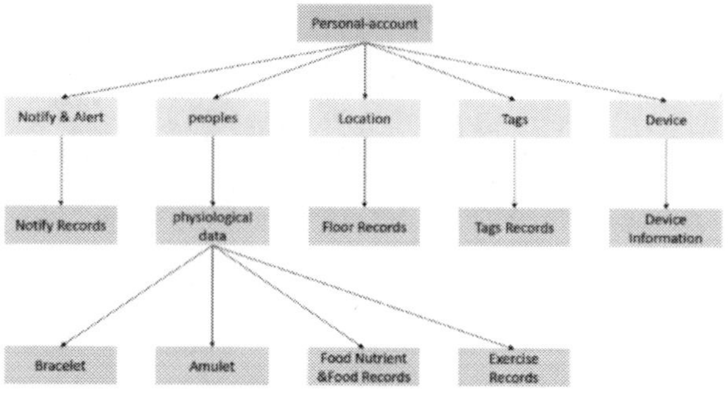


Figure 9. Data model.

In this experiment, we obtained a total of 3,344,651 data records from the Bluetooth bracelet; 1,528,487 records for posture changes of the amulets; 13 records for abnormal data; 20 records for notifications from Google Home Nest Mini; 105 records for food Photo records; 210 nutrient records; 11489 tag records, 10354 equipment records. In this study, we analyzed the subjects' physiological conditions, behavioral patterns, and eating habits. Due to the limitations of the experimental site and experimental time, we could not collect relevant data on sleep and exercise habits. The full quantity chart is shown in Table 1.

TABLE I. 10-DAY EXPERIMENT DATA VOLUME

| **Data field** | ***Numbers of row data*** |
|---|---|
| Bluetooth bracelet records | 3344651 |
| Bluetooth amulet records | 1528487 |
| Tag-records | 11489 |
| Device-records | 10354 |
| Data exception records | 13 |
| Speaker notification records | 20 |
| Diet photo records | 105 |
| Nutrient records | 210 |

### C. *Physiological Status Analysis*

This part focuses on the use and analysis of the data of the bracelet. You can view the changes of the subject's physiological data and other related information in time through the display of the webpage, as shown in Figure 10, or you can extract the history stored in the database by extracting. The data is analyzed, and a personalized physiological model is analyzed for different people so that the data can be effectively and correctly processed in the subsequent analysis of behavior and sleep.

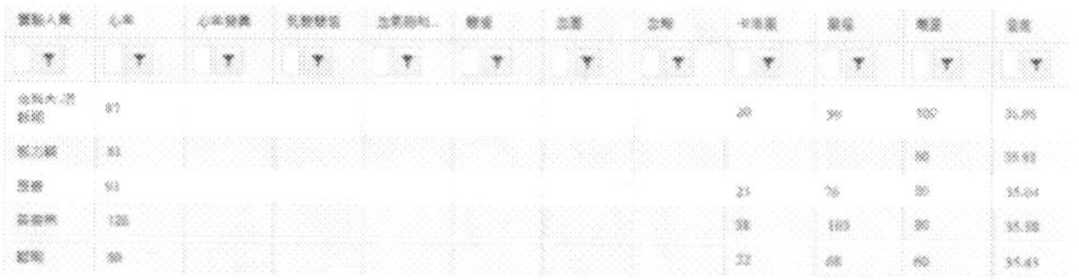

Figure 10. Web page real-time data update.

One subject was selected from the experiment for detailed analysis, as shown in Figure 11 and Figure 12, which are the daily records of heartbeat and body temperature, respectively. It can be found that most of the heartbeat and body temperature fall within the stable range, but some abnormal values can be seen from the figure. When the abnormal value occurs, the speaker will broadcast a reminder immediately.

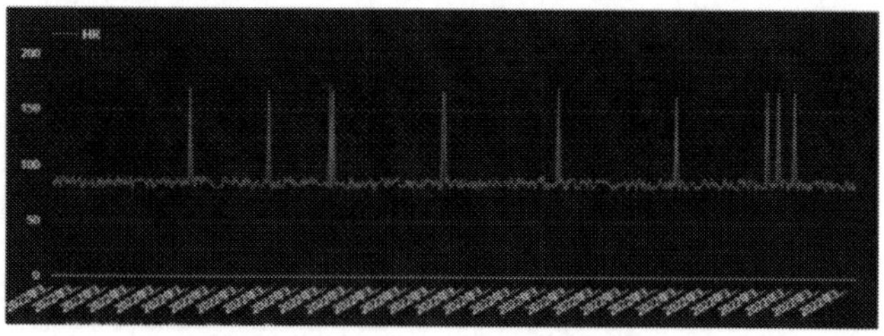


Figure 11. Heart rhythm model diagram1.

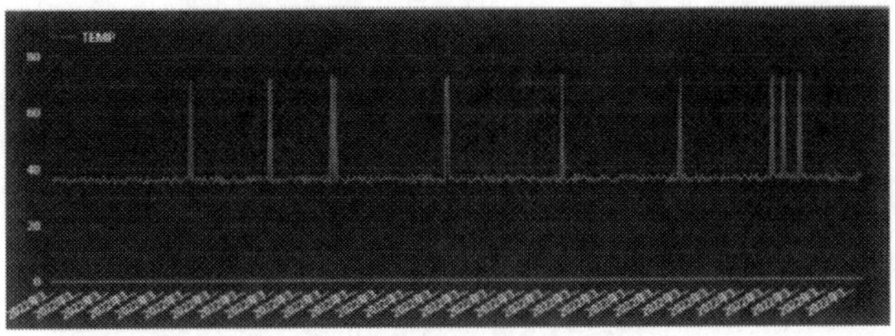


Figure 12. Heart rhythm model diagram2.

By recording and comparing each person's physiological information over a long period of time, a personalized physiological information model can be constructed, which can be clearly recorded when sleeping, walking, and sitting in different situations. Due to the influence of experimental environmental factors, the analysis and simulation in sleep and other situations cannot be more effectively displayed, otherwise, the importance of physiological information in this study can be more clearly realized.

## D. *Behavior Pattern Analysis*

Through the changes of positioning and posture, it can be very accurate to judge where the user is, what to do, and how long it takes, and through the daily record, the daily behavior of the user can be analyzed, as shown in Figure 13. It is a graph of the number of records, the number of experimenters, state changes, etc.

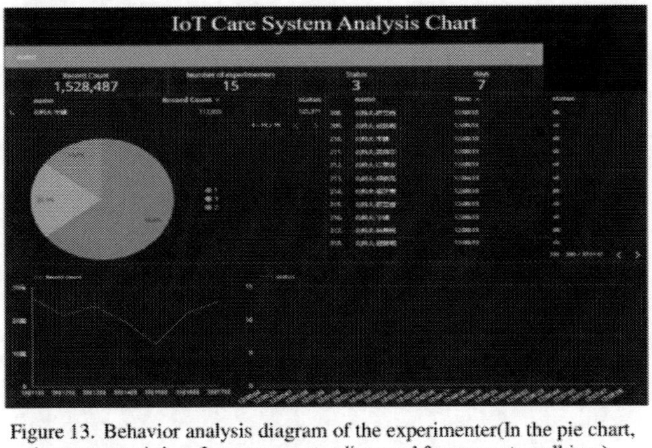


Figure 13. Behavior analysis diagram of the experimenter(In the pie chart, 1 represents sitting, 2 represents standing, and 3 represents walking.).

In the following, we select the posture changes of 2 subjects for further analysis and discussion. As shown in Figure 14, this is the posture record chart for a week, with Person 1 on the left and Person 5 on the right. It can be found from the figure, the proportion of 2 subjects sitting is extremely high, as high as 85.2 % and 77.9 %, respectively, there is a risk of sedentary, and it can be seen that the activity ratio of Person 5 is higher than that of Person 1, so they will use the speaker when they are sitting for a long time. Proper reminders are given to encourage increased activity.

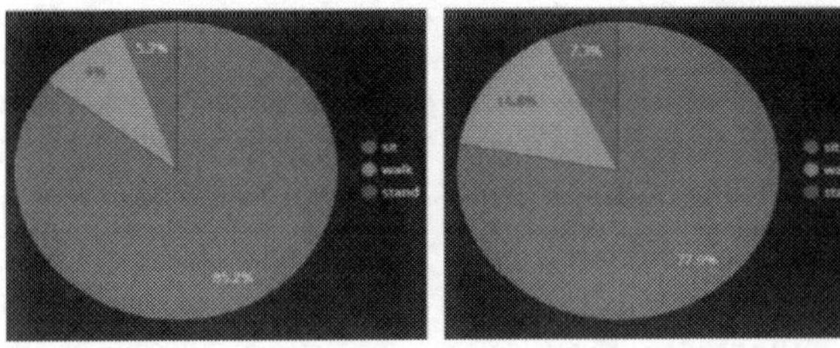


Figure 14. Attitude proportion pie chart(Person 1 on the left, Person 5 on the right).

Next, we observed the daily posture change curve of Person 5 within a week, and selected the curve changes of Day 2, Day 3, and Day 6 for discussion, as shown in Figure 15, Figure 16, and Figure 17. The vertical axis in the figure is the posture, 1 represents sitting, 2 represents standing, 3 represents walking, and the horizontal axis is the time axis. It can be found that the attitude changes models of Person 5 on Day 2 and Day 3 of the experiment are very similar, representing Person 5 in the week. In the afternoon of some days, there are similar itineraries and attitude changes, and these attitude models can be effectively recorded. Combined with the analysis of the position records, the activity position and activity itinerary can be further analyzed. Figure 17 obviously has a completely different itinerary from the other two days, which obviously means a new behavioral attitude model, which is completely recorded, which is conducive to longer-term experimental behavioral analysis.

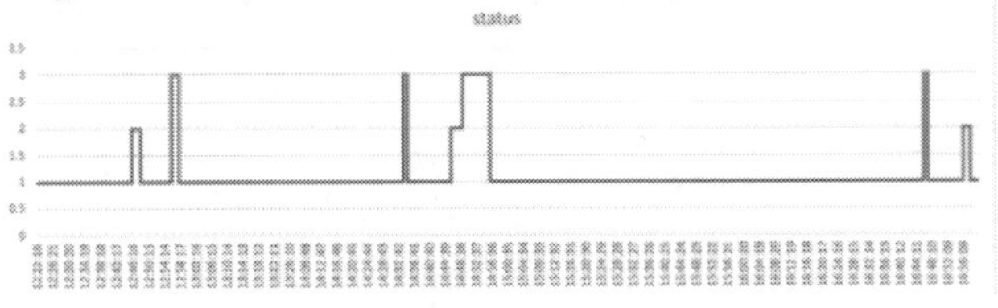


Figure 15. Attitude change model diagram1.

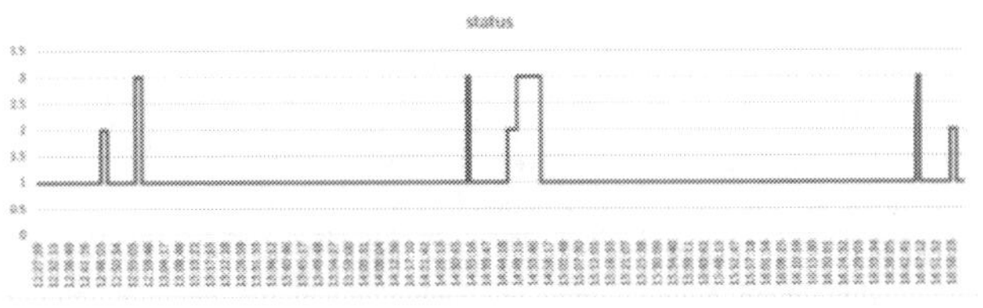


Figure 16. Attitude change model diagram2.

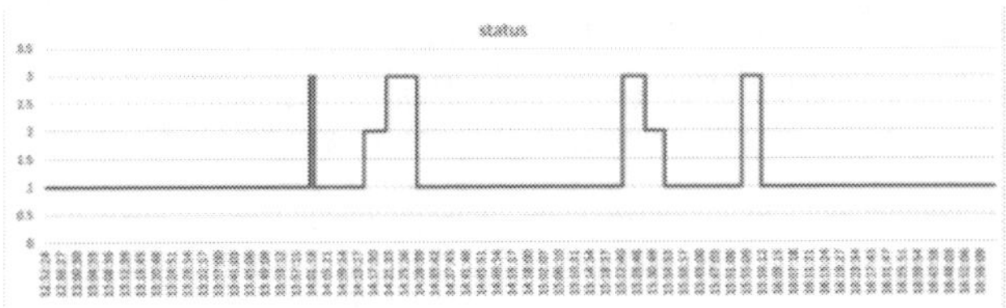


Figure 17. Attitude change model diagram3.

According to the above analysis content, in fact, the elderly's habits of getting up, going to the toilet, brushing teeth, eating, watching TV, sitting for a long time, standing for a long time, etc. can be accurately confirmed, and the behavior model can be recorded, and when abnormal activities occur. Notifications are made through speakers or LINE APP. Among them, sitting for a long time is a behavior that is easier to ignore and more likely to occur. Sedentary is related to cancer prevention in the elderly [10]. Therefore, the assistance of speakers is more needed to improve the elderly's health and bad habit.

*E. Diet Habit Analysis*

In order to clearly grasp the subjects' eating habits, we calculated the volume of food through the depth camera and combined deep learning to identify the type of food, and then grabbed and calculated the intake of nutrients, conducted a 7-day experiment, and recorded the daily lunch. Food content, a total of 105 records were collected and recorded in the database.

We selected the dietary contents of 2 subjects from the experimental data for comparison and analysis, and presented the data in the form of a table. We can clearly see the content and types of food they eat every day. The dietary contents are shown in Figure 18.

| | | | | |
|---|---|---|---|---|
| person1 | Day1 | Day2 | Day3 | Day4 |
| | rice:254g<br>cabbage:105g<br>steamed egg: 145g | noodle | noodle | noodle |
| | Day5 | Day6 | Day7 | |
| | rice:251g<br>tofu:92g<br>steamed egg: 163g | noodle | rice:202g<br>tofu:105g<br>steamed egg: 140g | |
| person5 | Day1 | Day2 | Day3 | Day4 |
| | rice:254g<br>stir-fried vegetables:105g<br>tofu: 145g<br>brocoli:198g | fried rice:548g | rice:237g<br>stir-fried vegetables:165g<br>tofu: 95g<br>brocoli:125g | rice:237g<br>stir-fried vegetables:165g<br>tofu: 95g<br>brocoli:125g |
| | Day5 | Day6 | Day7 | |
| | rice:207g<br>stir-fried vegetables:146g<br>brocoli:204g | rice:237g<br>fish:165g<br>stir-fried vegetables:176g<br>tofu: 84g<br>steamed egg:68g | rice:237g<br>shrimp:65g<br>stir-fried vegetables:213g<br>tofu: 54g<br>brocoli:105g | |

Figure 18. Weekly food content chart(Description of Person 1 and Person 5).

Arrange the data into a bar chart, as shown in Figure 19, from the figure we can easily observe the food intake and the number of items for a week. Person 1 prefers pasta in the staple food category, while Person 2 prefers meals. Then it can be clearly seen that Person 1 eats less food, most of which are about 3, and the intake of vegetables is not enough. On the contrary, Person 5 is in the intake of food types is relatively sufficient, and the intake of various seafood, vegetables, beans, etc., the intake of nutrients is quite sufficient for Person 1.

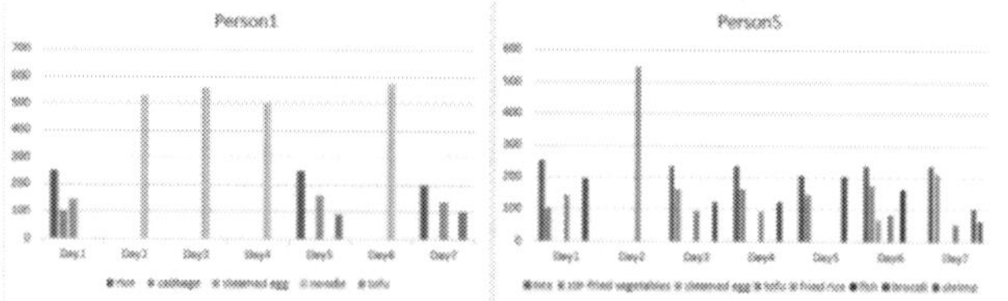


Figure 19. Histogram of weekly food intake.

Figure 20 shows the nutrient content per 100 grams of food, and the content is the type of food contained in Figure 19. It can be found that the cholesterol content of streamed eggs and shrimp is high, and the sodium content of noodles, fried rice, and shrimp is high, so you should pay attention to the intake of these foods containing higher nutrients.

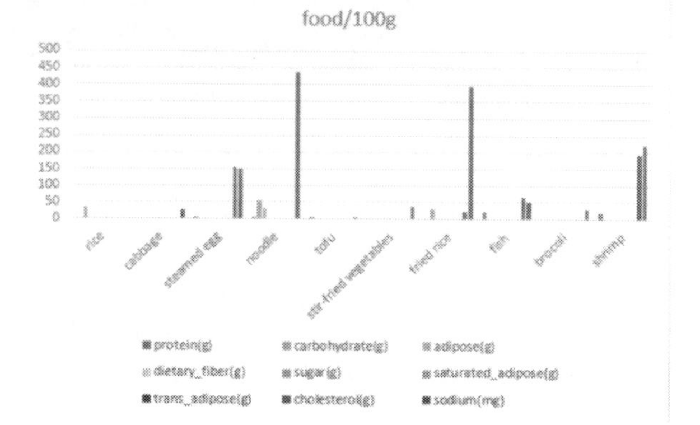


Figure 20. Histogram of nutrient content per 100 grams.

As shown in Figure 21, by observing the total nutrient intake of Person 1 and Person 5 for one week of lunch, it can be found that some situations are that Person 1 consumes too much sodium, which is much higher than the sodium intake of Person 5, and then uses the daily nutrients provided by [8]. By comparison, it can be more accurately found that the sodium intake of person 1 at noon is already close to the upper limit of the intake of three meals a week, so when the daily intake exceeds the standard, an immediate reminder will be given to inform you of the lack of those nutrients.

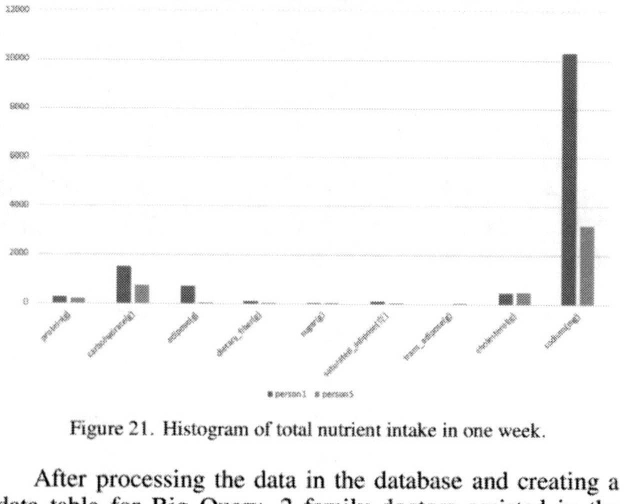

Figure 21. Histogram of total nutrient intake in one week.

After processing the data in the database and creating a data table for Big Query, 2 family doctors assisted in the analysis and verification of the data. With the help of doctors, we can have more accurate analysis results. Then, the analyzed data was discussed with the subjects. The subjects reported that there was a lack of dietary content, and when communicating with the Google Home Nest Mini, the speakers occasionally could not understand the user's words.

For this, we gave some responses, regarding the food image recognition ability, more image recognition is still needed and more diverse food models are established. Regarding the Google Home Nest Mini, it is necessary to strengthen the design of sentences and the use of different languages, and it can be modified in the program to match the relevant language supported by the speaker.

Finally, we effectively apply this intra-experimental result in real life. By observing the nutrient intake status every day, it is possible to immediately suggest the part of the nutrient intake that is insufficient on the next day, but the daily observation may not be able to understand the number of missing nutrients. The nutrients that the elder lack is very large, especially in the intake of some vitamins A, D, E, etc., it is easy to be ignored, so with the real-time reminder of Google Home Nest Mini, and the required food intake. It is recommended to gradually improve the user's eating habits, and also through the analysis of records, to know the relationship between the elderly's eating habits and physical conditions, to more effectively improve the elderly's bad eating habits, and to build a personalized diet model.

## V. CONCLUSION AND FUTURE WORK

This research is devoted to making a personal physiological data analysis system, completing data collection through the Internet of Things network combined with simple Bluetooth devices and cloud platform applications, and establishing a complete data analysis, through Google Home Nest Mini has established a good interaction with the elderly. Through daily data analysis and active push notifications, the elderly can instantly understand their own conditions and precautions. Through long-term data analysis, the following items can be effectively analyzed: Behavior patterns, eating habits, sleep quality, exercise habits, closely record and analyze the living conditions of the elderly, and through long-term analysis and suggestions, effectively improve the elderly's bad eating habits, and also use the Bluetooth wearable device. And the combination of the questionnaire content can more accurately analyze the sleep status and exercise habits of the elderly. These analysis contents can completely display the daily life pattern of the elderly and establish a personal model.

Although we have completely designed the entire care system and conducted some experiments and tests, the data volume and experimental environment of the experiment are still not perfect, and the data analysis can still be improved. In the future, we will continue to optimize the system and analyze and optimize it, and add the application of deep learning, so that the entire system can completely analyze and simulate various conditions of the elderly.

It is hoped that in the near future, a perfect personal digital avatar will be established through this system.

## REFERENCES


[1] H. Gu, Y. Diao, W. Liu and X. Zhang, "The Design of Smart Home Platform Based on Cloud Computing", International Conference on Electronic & Mechanical Engineering and Information Technology, 2011.

[2] T. Mondal, S. Jayadeva, R. Pani, M. Subramanian, Ashokkumar and B. k. Sumana, "E marketing strategy in health care using IoT and Machine Learning", DOI: 10.1016/j.matpr.2021.11.417,pp. 2087-2091, 2021.

[3] D. C. Yacchirema, D. Sarabia-JáCome, C. E. Palau and M. Esteve, "A Smart System for Sleep Monitoring by Integrating IoT With Big Data Analytics" in *IEEE Access*, vol. 6, pp. 35988-36001, 2018.

[4] Geriatric Depression Scale, Keelung City Government, 2017.

[5] Pittsburgh sleep quality index(PSQI), University of Pittsburgh, 1989.

[6] O. Dawelbeit and R. McCrindle, "Efficient Dictionary Compression for Processing RDF Big Data Using Google BigQuery" 2016 IEEE Global Communications Conference (GLOBECOM), Washington, DC, 2016, pp. 1-6, doi: 10.1109/GLCOM.2016.7841775.

[7] Dietary Reference Intakes(DRIs), Taiwan Health Promotion Administration, Ministry of Health and Welfare, 2021

[8] J. T. Huang, L. Y. Chang and H. C. Lin, "Implementation of IoT, wearable devices, Google Assistant and Google Cloud Platform for elderly home care system", in ICT4AWE21-PP-23, 2021.

[9] Z. Liu and J. Wang, "Associations of perceived role of exercise in cancer prevention with physical activity and sedentary behavior in older adults", in ScienceDirect Geriatric Nursing, volume 44, pp. 199-205, March–April 2022.

[10] M. Islam, A. Rahaman and R. Islam, "Development of Smart Healthcare Monitoring System in IoT Environment" , SN COMPUT. SCI. 1, 185, 2020.

[11] A. Goyal, H. S. kanyal, S. K. and R. Khan, "IoT based cloud network for smart health care using optimization algorithm" , Informatics in Medicine Unlocked, Volume 27, 100792, 2021.

[12] R. Iqbal, F. Doctor, B. More, S. Mahmud and U. Yousuf, "Big data analytics: Computational intelligence techniques and application areas" , Technological Forecasting and Social Change, Volume 153, 119253, April 2020.

# Gluco Coach—A Self-Management Application for Type 2 Diabetes Mellitus

User testing to understand comfort levels and sustained patient engagement

Way Kiat Bong
Department of Computer Science
OsloMet – Oslo Metropolitan University
Oslo, Norway
e-mail: wayki@oslomet.no

Kuthethur Sneha Jagannath Das, Felix Janszen
Inpaqt Technology Solutions B.V.
16, Marconistraat
Rotterdam, The Netherlands
e-mail: sneha.das@inpaqt.com, felix.janszen@inpaqt.com

*Abstract*—**The increase in the number of patients with type 2 diabetes mellitus is a global concern. Using information and communications technologies, particularly mobile applications, to empower patients with remote self-management has shown promising results. However, there is still a lack of evidence regarding an app content's contribution to making sustained impacts on the users' lifestyle behaviors and hence health. To make a positive health change through such a self-management app, the users are expected to be engaged in using the app. This engagement has to be sustained for a long period of time to reflect on both lifestyle behavioral and health changes. In our project, we developed Gluco Coach, a type 2 diabetes mellitus self-management app focusing on supporting type 2 diabetes mellitus patients to achieve and sustain healthy lifestyle behaviors. This paper presents work on understanding type 2 diabetes mellitus patients' comfort levels and interest in using Gluco Coach. User testing were conducted, and the findings showed potential in Gluco Coach and resulted in a list of lessons learned concerning design that aims to provide high usability and functionality, positive user experience, and thus sustained patient engagement.**

***Keywords: type 2 diabetes mellitus; self-management; self-monitoring; patient-centered design.***

## I. INTRODUCTION

The prevalence of chronic conditions is increasing globally, and it affects the quality of life of many individuals; type 2 diabetes mellitus (T2DM) is one of these chronic conditions [1][2]. The presence of T2DM is associated with various other comorbidities, such as obesity, cardiovascular issues, and renal issues [2]. It is estimated that by 2040, the number of people with T2DM will increase to 642 million globally [3]. The blood glucose level (BGL) of a T2DM patient rises due to insulin resistance or the ineffective use of insulin. Hence, in T2DM care, the goal is to maintain the BGL within the target range. The crux of T2DM care is remote self-management of lifestyle behaviors from the patient's side [4].

Growth in information and communications technologies (ICT) has led to the development of mobile health (mHealth) solutions, which predominantly consist of smartphone applications (apps) and wearables. Studies have shown the benefits of mHealth in chronic disease management [5][6]. Through mHealth, virtual coaching can be implemented in aiding the self-management of lifestyle behaviors. Many apps focusing on T2DM care are commercially available on the Apple App Store and the Google Play Store. These apps have the features of blood sugar tracking, reminders for meals, tracking calories burned, and T2DM care information. These are some essential features required for optimal T2DM self-management.

A popular definition of user engagement focuses on the aspect of the users' experiences with a technology [7]. Studies have shown that positive user engagement is a prerequisite for achieving positive health effectiveness [8][9]. However, while these apps are initially engaging for users, the aspect of sustained user engagement has not been thoroughly addressed [8][9]. Bringing about a positive health behavior change through an mHealth intervention requires users to engage with the interventions. This engagement needs to be sustained for a long period to reflect on the patients' behavioral and lifestyle changes. In addition, there is insufficient thorough information available about app content such as lifestyle behaviors along with evidence concerning their efficacy [10]. Studies have also emphasized the point that these apps still lack the clinical focus that a T2DM app requires [10][11]. Some examples of these apps are My Sugr Diabetes Logbook, Diabetes Pal, Diabetes Connect, Sugr Sense, and Health2Sync [10].

To address these knowledge gaps, in our project, we developed Gluco Coach, a T2DM self-management app focusing on supporting T2DM patients to achieve healthy lifestyle behaviors. The app is targeted to be science- and evidence-based concerning the aim of sustaining the patients' user engagements through personalization, which is lacking in current T2DM apps [10]. The personalization factor will be implemented using artificial intelligence (AI), which forms the crux of Gluco Coach. The evidence-based approach will lead the way to proving the clinical efficacy of the app. This position paper describes user testing conducted among T2DM patients to understand their comfort levels and interest in using Gluco Coach. These user testing were conducted as one of the first steps to understand patient engagement in relation to using a self-management app. In our earlier study, the CeHReS roadmap methodology was used throughout the development of Gluco Coach 1.0, where expert stakeholder sessions, think-aloud sessions, and a questionnaire study with T2DM patients were implemented to understand user engagement [12].

## II. GLUCO COACH

Gluco Coach is currently under development as a science- and evidence-based intervention. Aspects of T2DM physiology, along with behavior science, have been integrated into the app design to bring about the core feature of self-management. Pursuit of regular physical activity, a healthy diet, and adherence to prescribed medicines or insulin are integral to optimal T2DM care and can be understood as healthy lifestyle behaviors for the patients [11][13]. These features, along with T2DM care awareness messages, are the core functions of the app.

In addition, features such as personalized goal setting, self-monitoring, providing feedback, reminders, cues, and suggestions were incorporated into the app as virtual health coaching. The aim was to have users achieve healthy behavior changes. To implement this virtual health coaching, various behavioral science theories, such as goal-setting theory, the health belief model, the information-motivation behavior skills model, and protection motivation theory, were incorporated in the app design [14][15][16][17]. These theories were chosen as they involve personalized goal setting, adapted to the user [14], and provide information about T2DM care and the user's individual progress to motivate them to pursue healthy behaviors [15][16][17].

Figure 1 illustrates the main functionalities of Gluco Coach. Users can set personalized goals in terms of step count and diet in terms of calorie consumption and provide feedback on goal achievements in the form of visuals such as pie charts. The component of goal setting comes from goal-setting theory, while the components of self-monitoring and feedback come from the health belief model, the information-motivation behavior skills model, and protection motivation theory. The component of providing general T2DM care information also arises from these theories.

## III. METHODOLOGY

In this study, user testing with four T2DM patients (P1–P4) were conducted. They were recruited by convenience sampling because they were easily accessible. Their demographic information is summarized in Table 1. First, they were briefed about the project. Second, they were asked to provide their consent prior to participating in the user testing. The user testing consisted of providing their demographic background, performing a series of testing tasks, and answering a System Usability Scale (SUS) questionnaire. To avoid participants feeling exhausted, the testing tasks had to be prioritized, which were as follows:

1) Sign up (using a username and one-time code).
2) Enter physiological information, i.e., weight, height, blood pressure, BGL, and presence of other comorbidities. The information was based on what the participants knew previously, e.g., from their last health check, last self-measurement at home, etc.
3) Type in a medicine name and time for intake (dosage was not required).
4) Change the medication time for intake reminders.
5) Delete the medicine.
6) Inspect the navigation function.
7) Inspect the home page.
8) Insert the goal for the step count.
9) Log diet.
10) Investigate the T2DM care function.

TABLE I. DEMOGRAPHIC INFORMATION OF ALL PARTICIPANTS

| | **Age (years)** | **Gender** | **Self-rated ICT skills (1 is *very bad* and 10 is *very good*)** | **Highest education obtained** |
|---|---|---|---|---|
| P1 | 56 | Male | 2 | High school |
| P2 | 34 | Male | 9 | Master's |
| P3 | 58 | Female | 4.5 | High school |
| P4 | 43 | Male | 8 | Bachelor's |

When the participants were performing the tasks, they were observed and their actions were clarified, if required. They were also asked for their opinions on the main functions after completing the testing tasks. Each user testing lasted around one hour.

## IV. RESULTS

The participants could complete most of the testing tasks without much guidance. The two tasks that needed the most help were tasks 2 and 9. The reason P1 and P3 struggled with task 2 was because they did not understand some terms, such as "mmol/L" (millimoles per liter), a unit of BGL (see leftmost panel of Figure 1) and "Hemoglobin A1c" (Hb1Ac), a test for diabetes giving average (avg) BGL over the past two to three months. For task 9, the issues were due to finding a matching food name and amount. The function could only now offer to insert and search for a set of meals

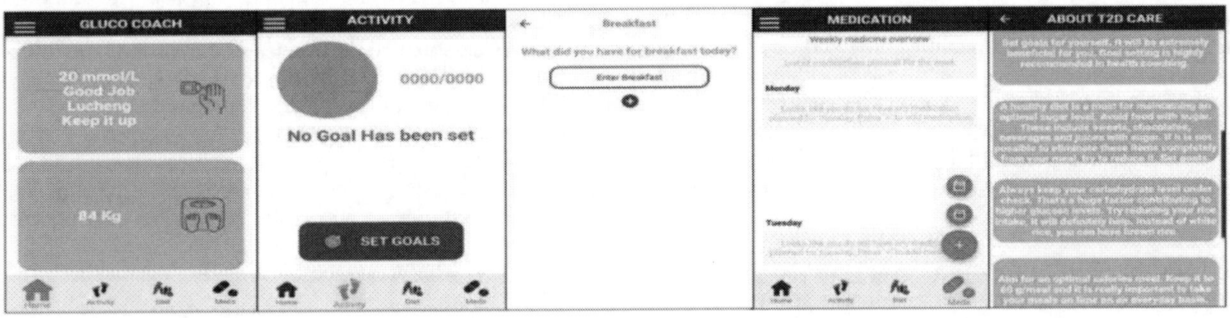


Figure 1. Gluco Coach. From left to right: home page, goal for step count, log diets, log medication intake, and T2DM care with awareness messages

with predefined amounts of food.

The participants managed to use the medication taking functions. They liked that the time for taking the medication could be set for each medicine and all participants appreciated the reminder function. They wanted it to be like an alarm, which could repeat to remind them to take the medicines, or the reminder could stay on the screen until the users acknowledged that they had taken the medicines. Only P4 perceived the logging diet function as useful to her. P1 and P3 expressed that they were not particularly interested in logging their diet. Instead, they preferred suggestions or to be instructed on what they should eat less or more of. They wanted to be guided to make better choices, rather than purely just logging the diet. Concerning the T2DM care function, all participants agreed that the information was useful. However, there was too much text; hence, it appeared boring and was not very attractive to them. They suggested having small tips provided to them one at a time, and the tips should be made based on their data.

TABLE II. SUMMARY OF AVG SCORE FOR SUS SURVEY

| **SUS statements (to rate from 1 to 5; 1 = *strongly disagree*, 2 = *disagree*, 3 = *neutral*, 4 = *agree*, 5 = *strongly agree*)** | **Avg** |
|---|---|
| 1. I think that I would like to use this system frequently. | 2.25 |
| 2. I found the system unnecessarily complex. | 3.25 |
| 3. I thought the system was easy to use. | 2.25 |
| 4. I think that I would need the support of a technical person to be able to use this system. | 3 |
| 5. I found the various functions in this system were well integrated. | 3.25 |
| 6. I thought there was too much inconsistency in this system. | 3.75 |
| 7. I would imagine that most people would learn to use this system very quickly. | 2.5 |
| 8. I found the system very cumbersome to use. | 2.5 |
| 9. I felt very confident using the system. | 3.5 |
| 10. I needed to learn a lot of things before I could get going with this system. | 2 |

Lastly, SUS results are presented in Table 2. Based on the avg scores, the participants perceived Gluco Coach as slightly complicated (statements 2, 3, and 8) due to inconsistency (statement 6). After clarifying with them, it was found that they wanted a more consistent and similar design for inserting meals and medicines. Both the placement and color of the buttons for adding meals and medicines were different (see Figure 1). The appearance of headings for some pages was not consistent, either. For example, the adding breakfast page in Figure 1 has a different heading from the other pages. Despite these issues, they agreed that the functionalities in Gluco Coach were well integrated (statement 5), which was aligned with their feedback when performing testing tasks. The participants saw potential in Gluco Coach as a self-management app for T2DM patients. Their SUS results indicate that they themselves have confidence in using it (statements 9 and 10).

## V. DISCUSSION

User testing with four participants for the first working prototype of Gluco Coach was conducted as one of the first steps in examining its design concerning usability, functionality, and user engagement. The findings indicate that Gluco Coach as a self-management T2DM app has the potential to sustain user engagement. However, some issues have to be addressed accordingly. Previous studies have indicated the importance of having features of reminders, personalized suggestions, feedback, and self-monitoring, which form a part of health coaching [18][19]. The diet function in Gluco Coach consists of a regular diary entry theme for logging food. From the user feedback, the feature of food suggestion would be much appreciated in addition to a more user-friendly way of logging diets. A personalized food suggestion that is relevant to the user can contribute to enhancing user engagement. Commercialized apps such as MyNetDiary [20] and HealthifyMe [21] have the feature of personalized food suggestions. These features can be implemented based on the person's BGL variation in a day, preferences, and allergies [11][22].

The feature of T2DM care information with awareness messages was appreciated, although the participants would prefer to have less text. To achieve this, coaching and awareness messages should be more personalized, like the above-mentioned personalized food suggestions, with tips to increase physical activities, for example. Personalization needs to be a pervasive feature of the app, as it supports user engagement [9]. These features can be implemented through AI with developing a context-aware feature, which can be utilized and implemented in the further development of Gluco Coach, since it will help in developing a more personalized and interactive user interface [23].

In task 2, it was observed that terms such as "mmol/L" and "Hb1Ac" were not easily understood by some participants. To alleviate such concerns, these terms can be explained to the users as a part of T2DM care and education in an interactive format. More visual and interactive forms of representation can be explored, such as the use of avatars to impart a more intuitive and fun way of raising awareness [24]. Lastly, the medication system of the app was well received by the participants. The idea of reminders was appreciated, and the participants suggested optimizing the reminders further in a more solid form. For instance, they should persistently appear on the screen until the users have taken the medicine and logged their medication taking.

Through user testing, a list of lessons learned have been compiled concerning design that will sustain user engagement among Gluco Coach users. They are summarized as follows, and they will be used to guide the further development of Gluco Coach in addition to suggestions from other relevant work [11][13][24]:

1) Offer personalized messages in the form of small tips that cover the three main areas of lifestyle, i.e., physical activity, diet, and medication adherence.
2) Provide necessary guidance or interactive education regarding T2DM.
3) Have a consistent design across pages in the app.
4) Make reminders more solid, e.g., persistent alarms and strong visuals on the app screen.
5) Suggest what to eat, besides logging the meals.

6) Offer a more user-friendly way to log diets, e.g., taking photos.

This study's major limitation is the small number of participants, and hence lacking diversity in the demographic backgrounds. Therefore, they do not represent the entire user group of T2DM patients. However, findings from the testing of four users can almost be sufficient to identify usability issues at this stage, as suggested by Nielsen [25].

## VI. CONCLUSION AND FUTURE WORK

This paper demonstrates the ongoing work of developing a self-management app, Gluco Coach, for T2DM patients. The app is targeted to be science- and evidence-based regarding the sustainability of patients' user engagement. User testing with four TD2M patients were conducted concerning their comfort level and interest in using the app. The findings indicate the potential of Gluco Coach despite some usability and functionality issues. Design recommendations were gathered as a list of lessons learned.

This study is one of the first steps toward understanding patient engagement. When conducting these user testing, the focus was on designs that provide good user experience and therefore sustained user engagement. Hence, our future work will include improving Gluco Coach based on the feedback from participants and lessons learned, and conducting user testing and a longitudinal study with more T2DM patients with diverse backgrounds to probe their user engagement.

## ACKNOWLEDGMENTS

We would like to thank the participants for their participation and willingness to contribute to this study, and also Yuan Jing Li from OsloMet for data collection, and Boris Mogilevski and Sumeet Kumar S Mulay from Inpaqt for their contribution in the development of Gluco Coach.

## REFERENCES

[1] M. A. B. Khan, M. J. Hashim, J. K. King, R. D. Govender, H. Mustafa, and J. Al Kaabi, "Epidemiology of type 2 diabetes–global burden of disease and forecasted trends," J. Epidemiol. Glob. Health, vol. 10, no. 1, pp. 107, 2020.

[2] M. C. Adriaanse, H. W. Drewes, I. Van Der Heide, J. N. Struijs, and C. A. Baan, "The impact of comorbid chronic conditions on quality of life in type 2 diabetes patients," Qual. Life Res., vol. 25, no. 1, pp. 175–182, 2016.

[3] World Health Organization, "Global Report on Diabetes," Geneva, 2016. ISBN: 978 92 4 156525 7

[4] M. Sina, J. Graffy, and D. Simmons, "Associations between barriers to self-care and diabetes complications among patients with type 2 diabetes," Diabetes Res. Clin. Pract., vol. 141, pp. 126–131, 2018.

[5] G. Maresca, M. C. De Cola, S. Caliri, R. De Luca, A. Manuli, et al., "Moving towards novel multidisciplinary approaches for improving elderly quality of life: the emerging role of telemedicine in Sicily," J. Telemed. Telecare., vol. 25, no. 5, pp. 318–324, 2019.

[6] K. Fan, and Y. Zhao, "Mobile health technology: a novel tool in chronic disease management," Intell. Med., vol. 12, no. 3, pp. 467, 2021.

[7] H. O'Brien, and P. Cairns, "An empirical evaluation of the User Engagement Scale (UES) in online news environments," Inf. Process. Manag., vol. 51, no. 4, pp. 413–427, 2015.

[8] L. A. Nelson, T. D. Coston, A. L. Cherrington, and C. Y. Osborn, "Patterns of user engagement with mobile-and web-delivered self-care interventions for adults with T2DM: a review of the literature," Curr. Diab. Rep., vol. 16, no. 7, pp. 1–20, 2016.

[9] L. Yardley, B. J. Spring, H. Riper, L. G. Morrison, D. H. Crane, et al., "Understanding and promoting effective engagement with digital behavior change interventions," Am. J. Prev. Med., vol. 51, no. 5, pp. 833–842, 2016.

[10] S. Izahar, Q. Y. Lean, M. A. Hameed, M. K. Murugiah, R. P. Patel, et al., "Content analysis of mobile health applications on diabetes mellitus," Front. Endocrinol., vol. 8, pp. 318, 2017.

[11] N. den Braber, M. M. Vollenbroek-Hutten, M. M. Oosterwijk, C. M. Gant, I. J. Hagedoorn, et al., "Requirements of an application to monitor diet, physical activity and glucose values in patients with type 2 diabetes: The diameter," Nutrients, vol. 11, no. 2, pp. 409, 2019.

[12] K. S. J. Das, and F. Janszen, "Identifying Sociodemographic Factors for a User Engaging Type 2 Diabetes Mellitus Mobile Self-Management Application," in 8th ICT4AWE, pp. 254–260, 2022.

[13] R. Shan, S. Sarkar, and S. S. Martin, "Digital health technology and mobile devices for the management of diabetes mellitus: state of the art," Diabetologia, vol. 62, no. 6, pp. 877–887, 2019.

[14] E. A. Locke, and G. P. Latham, "Building a practically useful theory of goal setting and task motivation: A 35-year odyssey," Am. Psychol., vol. 57, no. 9, pp. 705, 2002.

[15] N. K. Janz, and M. H. Becker, "The health belief model: A decade later," Health Educ. Q., vol. 11, no. 1, pp. 1–47, 1984.

[16] J. D. Fisher, and W. A. Fisher, "Changing AIDS-risk behavior," Psychol. Bull., vol. 111, no. 3, pp. 455, 1992.

[17] R. W. Rogers, "Cognitive and psychological processes in fear appeals and attitude change: A revised theory of protection motivation," Social psychophysiology: A sourcebook, pp. 153–176, 1983.

[18] I. Gupta, B. Di Eugenio, B. Ziebart, B. Liu, B. Gerber, et al., "Towards Building a Virtual Assistant Health Coach," in 2018 IEEE ICHI, 2018, pp. 419–421, doi: 10.1109/ICHI.2018.00081.

[19] S. Michie, M. Richardson, M. Johnston, C. Abraham, J. Francis, et al., "The behavior change technique taxonomy (v1) of 93 hierarchically clustered techniques: building an international consensus for the reporting of behavior change interventions," Ann. Behav. Med., vol.46(1), pp. 81–95, 2013.

[20] MyNetDiary. "MyNetDiary - Free Calorie Counter and Diet Assistant,"Available from: www.mynetdiary.com 2022.04.26.

[21] HealthifyMe. "HealthifyMe Indian Calorie Counter & Calorie Calculator App," Available from: www.healthifyme.com 2022.04.26.

[22] R. Z. Franco, R. Fallaize, J. A. Lovegrove, and F. Hwang, "Popular nutrition-related mobile apps: a feature assessment," JMIR Mhealth Uhealth, vol. 4, no. 3, pp. e5846, 2016.

[23] H. Op Den Akker, M. Cabrita, R. op den Akker, V. M. Jones, and H. J. Hermens, "Tailored motivational message generation: A model and practical framework for real-time physical activity coaching," J. Biomed. Inform., vol. 55, pp. 104–115, 2015.

[24] K. S. J. Das, T. Beinema, H. Op Den Akker, and H. Hermens, "Generation of Multi-Party Dialogues among Embodied Conversational Agents to Promote Active Living and Healthy Diet for Subjects Suffering from Type 2 Diabetes," In Proceedings of the 5th ICT4AWE, pp. 297–304, 2019.

[25] J. Nielsen, "How many test users in a usability study?" Available from: www.nngroup.com/articles/how-many-test-users 2017.

# Implementing and Learning to Use Video Meetings in Mental Health Hospital Departments

## Therapists' Experiences from Internal and External Meetings. A Qualitative Study

Monika K. Gullslett
Norwegian Centre for E-health Research
University Hospital of North Norway
Tromso, Norway
e-mail: monika.knudsen.gullslett@ehealthresearch.no

Elin Breivik
Norwegian Centre for E-health Research
University Hospital of North Norway
Tromso, Norway
e-mail: elin.breivik@ehealthresearch.no

Line Silsand
Norwegian Centre for E-health Research
University Hospital of North Norway
Tromso, Norway
e-mail: line.silsand@ehealthresearch.no

Etty R. Nilsen
Faculty of Health and Social Sciences
University of South-Eastern Norway
Porsgrunn, Norway
e-mail: etty.nilsen@usn.no

***Abstract*—Over several years, e-consultations and the use of Video Meetings (VMs) in the follow-up of patients in mental health services have become more and more common. During the winter and spring 2020, the pandemic accelerated and increased the use of VMs also in mental health services. The objective of this study has been to develop and advance comprehensive knowledge about therapists' use of VM in specialized mental health services. The therapists' external and internal collaboration is studied, as well as how their work is influenced. The study uses a qualitative approach, based on hermeneutic-phenomenological methodology. 33 interviews with therapists and management in three mental health hospital departments were carried out (using VMs) from March 2020 - February 2021. A semi-structured interview guide was used to encourage reflections on use of VMs. Overall, VMs as a communication tool is seen as efficient, in particular in remote areas. The results are presented in the following themes: lack of strategy for implementation and training, meeting structure and suitability of VMs for learning, security and safety measures pertaining to physical context, and managers' facilitation of VMs.**

***Keywords-implementation; organizational learning; mental health service; therapist; digital meetings; video meetings.***

## I. INTRODUCTION

Opportunities for new work methods have emerged for healthcare personnel during the pandemic and have caused a rapid increase in the use of video meetings (VMs) both in consultations with patients and with collaboration partners within mental health. Traditionally, therapists in mental health work in collaboration with their patients and collaborating partners through physical meetings. The focus of this article is the experiences of therapists in mental health hospitals of the use of VMs in meetings with collaboration partners.

Treatment and consultation via VMs represent a range of challenges for both therapists, patients and other partners involved [1]. We study the impact of using VMs through three different theoretical lenses: organizational learning, virtual communities of practice, and, finally, implementation of technology in organizations.

Organizational learning can be seen as collective processes of learning and knowledge sharing, and the focus is on the work situation as a shared context, and on the social processes in this context [2]. The shared context represents an opportunity for learning and knowledge sharing. A shared context facilitates learning, since knowledge, and especially tacit knowledge, cannot be counted on as flowing freely between the members of an organization if they do not interact or see each other work [3].

In the literature, shared contexts and situations are referred to as communities of practice [4], organizational spaces [5], and learning spaces [6]. The communities will vary along dimensions such as type of workplace and occupational group [7]. The increased use of technology (here VMs) has actualized the discussion of physical space vs. virtual spaces for the purpose of learning and sharing knowledge, and some studies pre-covid have investigated virtual communities of practice [8][9].

Where communities of practice are seen as emergent based on interest, and often informal, in the work environment, virtual communities of practice require initiative. They rarely have the feature that face-to-face communities of practice have, in that employees can bump into each other, or take an initiative with low effort and planning – like contacting the colleague next door.

Implementation of technology generally initiates a change process in organizations and has the potential to alter the way we work and the power relations in an organization. However, many change initiatives fail due to poor preparations and lack of training, and are slow to be implemented [10]. During the last two years, the pandemic accelerated and increased the use of VMs, also in mental health services [1]. Previous studies show that training is one of the situational factors that can hinder intention to use technology [11]. The pandemic has led to extensive use of VMs in various contexts. When lockdowns, social distancing and working from home due to COVID-19 were introduced, VMs were used to a large extent for all types of meetings, with or without patients, as well as within or between organizations. Over the last two decades a high number of studies has been published that identify barriers and opportunities for the implementation of VMs in health organizations, and lately studies that address the use of VMs during the pandemic. Practically, and in terms of saving time and being able to gather people from various organizations in the same meeting, the VMs are popular and considered efficient, in particular in remote areas.

Still, there is a knowledge gap to fill, particularly concerning the interaction and organizational learning when health professionals collaborate by using video technology. The aim of this study is to develop and advance comprehensive knowledge about therapists in specialized mental health hospitals and their experiences in using VMs in collaboration with different parts of the service, and how the use of VMs influences their interactions and organizational learning.

The research questions in this paper are 1) What are the barriers for interaction and organizational learning in internal and external VMs? and 2) How do the therapists experience both interorganizational interaction and interaction with collaboration partners in VMs?

The paper is organized as follows: In Section 2 we give an account of the Method and Study Design. This is followed by a presentation of the Results in Section 3. Section 4 contains the Discussion, and this is followed by the Conclusion in Section 5.

## II. METHOD AND STUDY DESIGN

### A. *Research design*

A qualitative and explorative study, using in-depth interviews, was conducted at three sites, all departments in different mental health hospitals. The study's methodological approach was based in the social sciences, using an iterative process between inductive and deductive strategy that aimed to uncover—and then interpret—knowledge about the actors in question [12]. This entailed exploring how the therapists experienced, understood, and created a context for using VMs in therapeutic meetings and interaction with colleagues and external collaborators. This perspective works well with the hermeneutic-phenomenological approach employed in our analysis; moreover, our choice of research strategy was integrated into the objectives of the study and the research questions under investigation [13][14]. The researchers' hermeneutic position will, even if the data gathering and analysis were done with a reflexive and open-minded view, affect the results based on the theoretical approach and researchers' preconceptions.

TABLE I. INFORMANTS IN THE STUDY - AN OVERVIEW

| Participants | Number interviewed | Interviews | Researcker |
|---|---|---|---|
| Hospital 1 | 14 | Video interviews | MKG |
| Hospital 2 | 13 | Video interviews | MKG |
| Hospital 3 | 6 | Video interviews | MKG |

In this context the concept "therapist" is used about mental health professionals, who are trained to provide treatments in different ways. There was diversity in age, gender, and professional background among the informants in all the hospitals: psychologists, psychiatrists, milieu therapists (nurses and social workers), all with at least three years of university education. The informants were 27-66 years at the time of the interviews.

### B. *Interviews*

Semi-structured interviews were conducted from late March 2020 to mid-February 2021. The interviews were carried out as VMs. The first author of the paper conducted all the interviews and opened each interview by asking the therapist to tell a story about when, how, and why they had implemented VMs in their mental health service for the first time. An interview guide was developed beforehand and sent out to all informants prior to the interview. When the Covid-19 restrictions were implemented, one of the recommendations for mental health workers was to follow up their patients by using VMs [15]. We had already planned a qualitative study at the hospital on different aspects regarding the organization and implementation of VMs during normal circumstances. When the societal lockdown occurred, we accelerated the process to investigate the therapists' experiences of being rushed into a large-scale implementation of VMs in the hospital environment. We sent a request for participation to the management at the first hospital on 20 March 2020. The management redistributed the request to everyone in the mental health departments, stating that participation should be given priority. The second hospital got the request in May 2020, and the third in October 2020. A similar procedure was followed at each site. We recruited a total of 33 employees from different disciplines and departments.

### C. *Analysis*

All interviews were recorded and transcribed verbatim. The transcriptions were undertaken by a professional firm just after the interviews were completed. To validate the content, the first author read all the transcriptions and compared them to the recorded interviews. The analysis was conducted through a reflexive, open-minded and abductive process, which enabled an intuitive understanding of the meaning of the text as a whole [16]. Following the initial in-depth reading of the interviews, the content was categorized and grouped together to identify important themes according to the research questions. The themes in the analysis emerged through an iterative process of reading and interpreting, to identify meaningful units [12][16]. All authors conducted the analysis and reflected on the findings together with the first author, who read the most central nodes coded in NVivo. All authors contributed to the writing of the paper's background, discussion sections, and its revisions.

*D. Ethical approval and considerations*

The study was approved in advance by the ethical committee (PVO), Helse Nord (project ID 2462). The participants were given both written and verbal information about the study, before agreeing to participate. The included informants sent their consent via mail to the first author, and these were stored without any connection to the gathered data material. The participants signed an informed consent form. The data are anonymized in the presentations.

*E. Authors' contributions*

All authors made significant contributions to the manuscript. The study was conceived by MKG, and was drafted in close cooperation with ERN, LS and EB. MKG collected data and MKG, LS and EB contributed to the analysis. The manuscript was written by MKG, ERN, LS and EB, and all authors read and approved the content of the final manuscript.

## III. RESULTS

The focus of the paper is the therapists' experiences with implementation and use of VMs, and, in particular, their opportunities for interaction with collaboration partners and learning when video technology sets premises for practice. In this study we focus on how the therapists experience their interorganizational interaction, as well as their interaction with external collaboration partners. Furthermore, we investigate barriers for communication, interaction, and organizational learning in VMs. The following main themes emerged from the analysis: A. Learning and training when implementing VMs in the organization (shared context), B. Content and context – barriers and drivers in VMs, and C. Culture and structure in VMs.

*A. Learning and training when implementing VMs in the organization (shared context)*

The informants in the study were asked about the training they received when the hospitals rushed into a large-scale implementation of VMs, due to the societal lockdown during the covid pandemic.

The informants distinguish between two types of training, where the first encompasses being able to use the video technology and to set up and connect to VMs. The other concerns how to interact both with internal and external partners in VMs. Informants from one of the organizations had been trained and had used VMs regularly during the last years prior to the pandemic. However, most of the informants had received training during test periods, for instance as part of a research project:

> "*We received training together with colleagues to test VMs for a period of time to learn how to use VM in connection with a research project. [...] It was up to the staff if they wanted to use it further...*".

All in all, few of the informants have used VMs in their professional practice prior to the pandemic. In some clinical units, super-users have been qualified and given the responsibility to train colleagues. In other units, training was often limited to support from colleagues. Independent of the type of training and previous experiences with VMs, the start-up phase was characterized by trials and errors. Several informants emphasized continuous training as important to become familiar with the VM tools. The sharing of experiences between colleagues was highlighted as essential in this process.

> "*Some of us colleagues got access to and training in use of VMs and we passed on training to the others. (...) We carried out training within the team, and we simply learned a little from each other. (...) Gradually it became more systematized, but at the very beginning of the pandemic, there was a lot of trial and error*".

Informants with different degrees of experience in using VMs, reported frequent technical problems as a main challenge. Technical problems are a common source of frustration and an obstacle for continuing use of VMs after the pandemic. These problems are time and energy consuming during meetings, as the following quote illustrates:

> "*You spend a lot of time and energy getting the technology and different solutions to work together; someone falls out, comes in, does not hear, the image freezes... So, it creates laughter and also a lot of frustration. It is unfortunate when addressing an important issue. If the digital tool had been effortless, then it would have been something that you should definitely continue with, because you otherwise spend a lot of time traveling or others have to come in here*".

This quote also illustrates the connection between technical problems and lack of technical proficiency on one hand, and the serious character of the content in the meetings on the other. This is a matter that we will return to further down.

*B. Content and context – barriers and drivers in VMs*

Apart from learning how to use the technology; another type of learning revolves around how VMs should be accomplished. The informants reflect on the need for training and structures that help to achieve a safe environment with meaningful and valuable discussions in VMs, and not only focus on efficiency.

*"Hmm... It´s like a voice in my head says no, VMs can never replace physical meetings. But I don't know if it's just because it's different to meeting digitally. Because we are taught to meet in physical rooms, right? We have never been trained in digital evaluation or digital assessment or digital treatment. It has never been a topic".*

Uncertainty related to context and content may be a barrier to using VMs. Nevertheless, training in how to communicate and interact in digital meetings has never been raised as an issue in most departments. An important issue raised by several of the informants' concerns safety and privacy when actors from different organizations connect and collaborate through VMs. The informants were specifically concerned with protecting patient privacy, being exposed to conversations about patients, often connected to persons present in the room who should not be part of the meeting. This is particularly important when sensitive patient issues are discussed, and particularly pertinent in a health care setting, as the following quote shows:

*"It turned out that there was another person in the room that I was not aware of being there. It was actually a bit uncomfortable. I thought I knew who was in the room ...".*

The quote below underscores this point further: using VMs for discussions of sensitive information requires that the organization has prepared physical rooms suitable for VMs and can be difficult to comply with when healthcare personnel work from home. This might appear as an oxymoron, since VMs are introduced to make the physical location less important.

*"Last week I had a meeting with a colleague when she was in the home office. It was the child welfare service and the two of us who had a meeting. Suddenly she started moving (the person moved to another room), and I heard someone coming in, so it was a bit of an awkward situation. And I thought that if I had been a patient, I would have had the feeling; who is it in the room now? It was very inappropriate."*

Many informants pointed out that it was difficult to address complicated issues, especially if there are many participants in the meeting. In-depth discussions about for instance professional matters. were by many perceived as less suitable for VMs.

*"It is not so easy to communicate when there are more than 5-6 people. Then it becomes difficult to communicate and discuss. But if a leader is only going to give information about things that are important for us to know, then it works with many in the same meeting, because then we mute and listen".*

VMs work well for giving and receiving information, as well as for addressing administrative matters. Several participants in this study shared that it is easier to plan meetings and ask for advice from collaborating partners, when meeting digitally. VMs demand less time when there is no need to travel to a physical meeting, and less time is used for small talk compared to physical meetings.

*"I think it (VMs) is very suited for information exchange. A bit also because there are so many participants and finding a time that suits everyone is difficult. I have many patient consultations during the day, so it is easier for me to find three quarters of an hour - an hour between consultations than having to travel to the school or the health centre".*

*C. Culture and structure in VMs*

Another issue brought up by the informants, was meeting culture and meeting structure when using VMs. Several informants found that VMs are well structured (by nature), and that they are allowed to speak without being interrupted. Furthermore, as VMs are often shorter and more to the point than physical meetings, it is easier to find time for them.

*"I feel like it is either it's working or not (VMs). It works very well with the persons you are used to working with, and who use the same technology as you. Then I think it works well".*

However, some of the informants address the need for a new and different awareness of each other when participating through VMs. The changed awareness referred to the above-mentioned structure in meetings, and that all participants get their time to speak during meetings, since it is difficult to observe body language through a video camera.

*"The challenge in VMs is that often everybody is talking at once. Because, when everybody is sitting in the room you see very quickly if someone is going to say something, but you do not do that [in VMs] until they have started talking. So, it becomes challenging to get the discussion to flow".*

In meetings with many participants, most turned off their cameras. This made it hard for the chair of the meeting to judge how the information was received. Some of the informants emphasized that it is challenging to exercise good leadership and to chair in VMs because of less feedback from the participants, compared to physical meetings. A concern is that it in the long run, the interaction between colleagues and collaborators will be affected, and there is a danger that the work environment will deteriorate.

*"(...) I miss the meetings where we all gather (...) I also believe that during the pandemic, many people has been "locked inside" their own little bubble and is busy with their own stuff. The consequence is, that we can act and do things without any influence from our colleagues or collaborators – and that is not a good thing".*

## IV. Discussion

The study has displayed a series of challenges for the use of VMs in mental health service meetings, internally and externally. These challenges pertain to implementation and learning, more particularly to

- Lack of strategy for implementation and training
- Meeting structure and suitability of VMs for learning
- Security and safety measures pertaining to physical context
- Managers' facilitation of VMs.

This study demonstrates that the experiences of the therapists' concern at least two areas: (1) learning the technological skill, and (2) learning how to carry out a successful VM. Successful is described by the informants as creating a safe environment with meaningful and valuable discussions. Implementation of technological tools for digital meetings, like the use of VMs are thoroughly researched within the IS field (see for instance [17][18]) and findings from the present study largely matches extant research on several matters: The strategy for training varied, from (1) training super-users, and expect the skills that the superusers acquire to trickle down, to (2) learning by doing. The continuous and iterative learning and support, both in the initial adoption phase and in the continuance, called for by research [19], was lacking, particularly in the initial phase. Despite the often lacking or flawed strategy for learning, the training phase of implementation within the organization offered an arena for rapid learning and co-creation amongst colleagues. The pandemic provided the urgency called for in change management [20].

Part of the implementation, as experienced by the therapists, was learning how to make the VMs work. As shown in previous studies, the technology in itself provides resistance in that it often does not work at all, or not according to its purpose [21]. In inter and intra organizational meetings, which additionally are interdisciplinary and cross-professional, and where the topic is, for instance, treatment plans for specific patients, thus highly sensitive, a digital field for learning across professions and organizations is created [4][22]. In VMs, however, the meeting structure, or lack thereof, becomes strongly visible since use of VMs demands more planning and structure than face-to-face meetings. Lack of structure appears to harm the interaction. The informants call for more structure, but a strict structure might on the other hand harm learning processes and development of innovative solutions.

The informants view the use of VMs as efficient, less costly, and easy to organize. However, when asked about the implementation and the content of the meetings, a number of barriers and challenges emerge from the data and there appears to be limitations as to the suitability for VMs in particular situations. Both inter-organizationally and with external actors, the meeting participants in this setting work interdisciplinary and inter-professionally. When the various professions and disciplines contribute with their knowledge, the meeting is an arena for learning and for sensemaking through interaction [23][24]. However, VMs differ from face-to-face meetings in several aspects, and require attention paid to structure and culture in a different manner for the participants to be heard and seen.

Several stages in the transfer from physical meetings to VMs appear to happen randomly and without a strategy, and this particularly affects the need to take security and safety measures. In the healthcare context, where they constantly handle sensitive information and personal data, two dimensions of safety emerged: (1) the digital safety and (2), ironically, the physical safety. During the pandemic, a part of the hospital's practice is moved to the private sphere: the homes. During treatment, it is the home of the patient in one end of the conversation and the home-office of the therapists in the other [1]. In this study, the therapists interact with each other and with other professional groups and individuals, within or outside their organization. Security and safety measures are similarly relevant since the conversations evolve around treatment of vulnerable individuals.

Encryption does not ensure what we have coined as *physical safety*, meaning securing the physical environment where the therapists are situated when s/he works from home. Ensuring the physical safety, for example who is present, is an organizational and managerial issue, and points to lack of facilitation and control on the management's part, in order to create a safe and evolving digital community.

VMs are not a static form of collaboration. On the contrary, the study shows that interaction through VMs take place in a new context, and organizational learning, including transfer of tacit knowledge, does not take place automatically, as the partners are not present in the same physical environment.

## V. Conclusion

In this study, the context, which is therapists in the mental hospital and their interaction, expands extant research and underscores the salience of context in the process of implementing and using technology. It demonstrates that training cannot be seen as 'one size fits all'. In this sense it is salient to raise the question of how each individual learn. Being digital and handling VMs successfully, is a skill that must be learned through training and use.

To further develop digital collaboration, in this case VMs, the organizations must focus on which organizational processes should be changed. e.g., whether it requires a change in workflow and whether changes in power relations occur.

Managers on all levels in the organizations must be involved in the implementation process with a clear strategy. To plan and perform useful VMs require managerial facilitation and considerations that are novel compared to regular physical meetings.

## ACKNOWLEDGMENT

This paper stems from the research project, 'Distance Follow-Up in the Specialist Health Care Service', funded by The Ministry of Health and Helse Nord. We acknowledge the support and assistance provided to us by the staff members of the mental health clinics in the hospitals while we conducted this research.

## REFERENCES

[1] M. K. Gullslett, E. Kristiansen, and E. R. Nilsen, "Therapists' Experience of Video Consultation in Specialized Mental Health Services During the COVID-19 Pandemic: Qualitative Interview Study," JMIR Hum Factors, 8(3), 2021.

[2] S. Newell, M. Robertson, H. Scarbrough, and J. Swann, "Managing knowledge work and innovation," Basingstoke: Palgrave Macmillan, X, 277 s, 2009.

[3] G. von Krogh, K. Ichijo, and I. Nonaka, "Enabling Knowledge Creation. How to Unlock the Mystery of Tacit Knowledge and Release the Power of Innovation," Oxford: Oxford University Press, 292, 2000.

[4] E. Wenger, "Communities of Practice. Learning, Meaning, and Identity. Learning in doing, social, cognitive and computational perspectives," Cambridge: Cambridge University Press, 1998.

[5] D. Yanow, "Talking about Practices: On Julian Orr's Talking About Machines," Organization Science, 27(12): p. 1743-1756, 2006.

[6] M. Thompson, Structural and Epistemic Parameters in Communities of Practice," Organization Science, 16(2): p. 151-164, 2005.

[7] K. Handley, T. Clark, R. Fincham, and A. Sturdy, "A space for reflection and learning? An investigation of physical, relational and existensial space in client-consultancy projects," in OLKC 2006. University of Warwick: Unpublished, 2006.

[8] M. A. Ergan, A. T. Vold, and E. R. Nilsen. "Virtual Communities of Practice - Experiences from VCoP," in

[9] Proceedings of the 15th European conference on knowledge management, 2014.

[10] S. Valenti, and S. Sutton, "Strengthening Virtual Communities of Practice (VCoPs): An Evidence-Based Approach," Journal of Education for Library and Information Science, 61(1): p. 106-125, 2020.

[11] H. K. Andreassen, L. E. Kjekshus, and A. Tjora, "Survival of the project: A case study of ICT innovation in health care," Social Science & Medicine, 132(0): p. 62-69, 2005.

[12] A. M. Fuglseth, and Ø. Sørebø, "The effects of technostress within the context of employee use of ICT," J Computers in Human Behavior, 4. 40: p. 161-170, 2014.

[13] N. Blaikie, "Approaches to social enquiry: Advancing knowledge," 2 ed., Cambridge: Polity Press, 2007.

[14] C. Pope, S. Ziebland, and N. Mays, "Qualitative research in health care: Analysing qualitative data," British Medical Journal, 320(7227): p. 114-116, 2000.

[15] S. Kvale, and S. Brinkmann, "Interviews: learning the craft of qualitative research interviewing," Los Angeles, Calif.: Sage. XVIII, 354 s. : ill., 2009.

[16] Directorate of eHealth, "Corona: How to get started with Video Consultations," Directorate of eHealth: Oslo, 2021.

[17] S. Kvale, and S. Brinkmann, «Interview: Det kvalitative forskningsinterview som håndværk,» Hans Reitzels Forlag, 2015.

[18] W. J. Orlikowski, and D. Robey, "Information technology and the structuring of organizations," Information systems research, 2(2): p. 143-169, 1991.

[19] K. Osmundsen, J. Iden, and B. Bygstad," Digital Transformation: Drivers, Success Factors, and Implications," in MCIS, 2018.

[20] Ø. Sørebø, and T.R. Eikebrokk, "Explaining IS continuance in environments where usage is mandatory," Computers in Human Behaviour, 2008.

[21] J. P. Kotter, "Leading change," Harvard Business Review Press, 2012.

[22] E. R. Nilsen, J. Dugstad, H. Eide, M. K. Gullslett, and T. Eide, "Exploring resistance to implementation of welfare technology in municipal healthcare services – a longitudinal case study," BMC Health Services Research, 16(1): p. 657, 2016.

[23] T. Eide, et al., "Trust-based service innovation of municipal home care. A longitudinal mixed methods study," In review: p. 1-27.

[24] T. Hernes, and O. Olsen, «Organisering i en verden i bevegelse. Organisering,» Oslo: Cappelen Damm akademisk, 2016.

[25] K. E. Weick, "Sensemaking in Organizations," Foundations for Organization Science, ed. D. Whetten. Vol. 1, Thousand Oaks: Sage. 229, 1995.

# Assistance Effect of an Evolved Heel-Raising Unit for Walking Disabilities

Haruki Baba, Akimasa Suzuki
Graduate School of Software and Information Science
Iwate Prefectural University
Takizawa, Japan
E-mail: g231s027@s.iwate-pu.ac.jp,
suzuki_a@iwate-pu.ac.jp

Yoshitoshi Murata
Research and Regional Cooperation Office
Iwate Prefectural University
Takizawa, Japan
E-mail: y-murata@iwate-pu.ac.jp

*Abstract*—**Elderly people and people with diseases such as hemiplegia often have a walking disability, which increases their risk of falling and suffering injuries. In walking, the angular velocities when raising the heel and swinging the toe forward are lower for elderly people than for healthy individuals, because of their lower muscle power. We previously developed a heel-raising unit based on a spring, but that unit lacked a mechanism to release the spring at the optimal timing, as reported in the International Journal on Advances in Life Sciences in 2020. We have since developed an evolved heel-raising unit that works when the heel starts to rise after the whole shoe has contacted the floor. Experimental results demonstrate that the evolved unit's assistance effect is better than that of the previous simple unit. We also report a correlation between body weight and the optimal spring stiffness, and we show that the evolved unit does not affect the individual's walking posture as much as the previous one did.**

***Keywords: walking disability; walking-assist unit; heel raising; walking posture; muscle power.***

## I. Introduction

As the percentage of elderly people in the world's population is increasing [1], the number of functionally impaired people, such as those with hemiplegia, will also increase. People with such diseases and very elderly people often have walking disabilities, which increase their risk of falling and injuring themselves [2].

Our previous study comparing the walking gaits of hemiplegia patients and healthy students showed that the angular velocities while raising the heel and swinging the toe forward were lower for people with walking disabilities than for healthy people. The reason was the lower muscle power of the hemiplegia patients in our study [3]. This suggests that assistance in raising the heel and swinging the toe forward could help people with walking disabilities to have a gait closer to normal.

Two kinds of foot prostheses, the Solid-Ankle Cushion-Heel (SACH) foot (e.g., 1D10, Ottobock, Germany [4]) and the energy storage and return (ESAR) foot (e.g., Vari-Flex, Össur, Iceland [5]), have enabled foot amputees to improve their gait to be close to normal. In the case of the SACH foot, a properly shaped wedge of cushioning material is built into the heel part of the foot prosthesis to absorb shock when the heel of the shoe contacts the floor and thus assist in raising the heel. The ESAR foot has a kind of leaf spring to absorb the contact shock and assist in raising the heel more effectively than the SACH foot does.

Inspired by the ESAR foot, we previously developed a heel-raising unit that used a spring [6]. The unit was very simple, as it comprised only a conical coil spring and a V-shaped attachment cover made of thin stainless steel, but it did not have a mechanism to release the spring at the optimal timing. As a result, it started to raise the heel as soon as the shoe's heel contacted the floor. Instead, the spring should ideally be released when the heel starts to rise in the gait cycle.

Hence, we have developed an evolved heel-raising unit in which the spring is released when the heel begins rising after the whole shoe has contacted the floor. Experimental results with the evolved unit demonstrated that its assistance effect was notably better than that of the previous unit, and that it did not affect the walking posture as much as the previous one did.

In Section II, we describe the differences in gait between a hemiplegia patient and a healthy person to clarify the required characteristics of an evolved heel-raising unit. Then, in Section III, we explain how the SACH foot and ESAR foot compensate for a lack of kicking power when raising the heel. Section IV describes the structure of the evolved heel-raising unit and the experimental results, before Section V concludes with a summary of the key points.

## II. Gait Differences Between Hemiplegia Patients and Healthy People

We analyzed the walking gait cycles of both unimpaired people and people with walking disabilities by using a wearable device (WD) and a Kinect to detect warning signs of falls [3]. Every patient with a walking disability in this experiment had hemiplegia and trained periodically at a rehabilitation facility. To estimate the kicking power and the change in angle between the foot and the floor, we experimentally measured the output data from an acceleration sensor and a gyroscope sensor in a WD mounted on the front of a shoe. A Sony SmartWatch 3 was used as the WD.

Following Tao et al. [7], we divided the normal walking gait cycle into eight phases as shown in Figure 1: (1) initial contact (heel-strike timing), (2) loading response, (3) mid-stance, (4) terminal stance (toe-off timing), (5) pre-swing, (6) initial swing, (7) mid-swing, and (8) terminal swing.

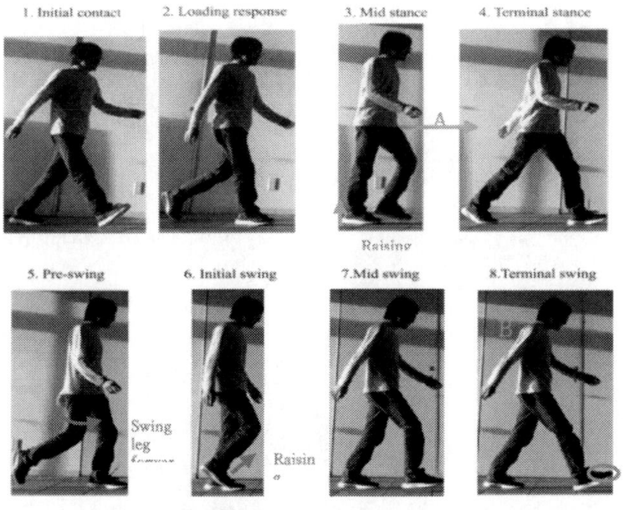


Figure 1. Normal walking gait cycle (right foot).

Figures 2 and 3 show examples of changes in the angular velocity, angle, and acceleration for a healthy participant and one with a walking disability, respectively, over the course of two steps. Each flat region (roughly in the center of each graph) represents the period when the shoe's entire sole touched the floor. The acceleration in this period is roughly 9.8 m/s$^2$ because the acceleration sensor still measures gravity. The maximum angular velocity at point A indicates the kicking power when raising the heel. This action occurs between the mid-stance and terminal stance, as shown in Figure 1. The minimum angle at point B indicates the toe angle to the floor at the terminal swing.

As seen in Figure 2, the lowest angular velocity at point A for the unimpaired participant was approximately 420°/s. In contrast, as seen in Figure 3, the highest angular velocity at A for the participant with a walking disability was only 250°/s. Thus, the latter participant clearly had a weaker kicking power than the healthy individual when raising the heel, which indicates a clear difference in gait.

Similarly, the highest angle at point B in Figure 2 for the unimpaired person was approximately −18°, whereas the lowest angle at B in Figure 3 for the participant with a walking disability was approximately −8°.

Tables I and II list the averages and standard deviations (SDs) of the measured angular velocity at point A and angle at point B. The angular velocity at A was clearly different between the unimpaired participants and those with walking disabilities. There was also a measurable difference between them in the angle at point B, although the ranges of those values sometimes overlapped.

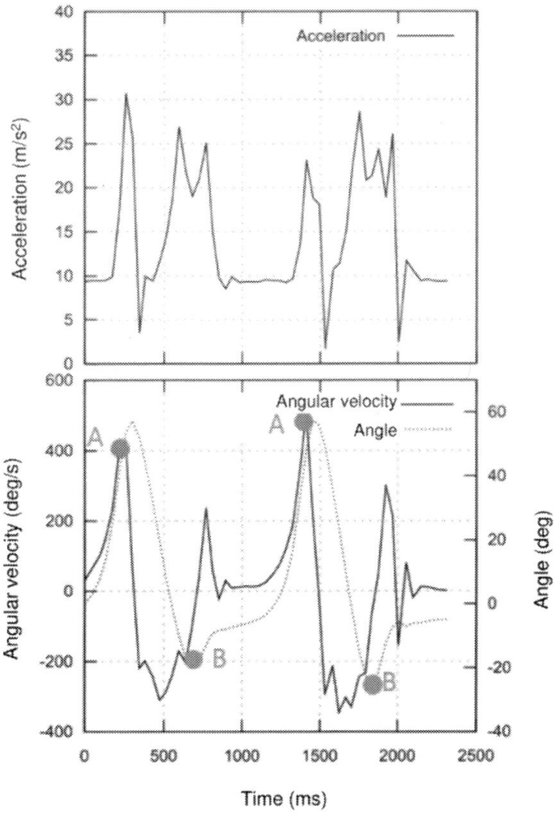


Figure 2. Examples of the angular velocity, angle, and acceleration for an unimpaired participant.

Thus, the individuals with a walking disability had a weaker kicking power when raising the heel than that of the healthy participants. They also had difficulty raising their toes during the terminal swing phase. Accordingly, if additional power could compensate for the difference in kicking power when raising the heel, an individual with a walking disability could walk with a gait closer to that of a healthy person.

TABLE I. ANGLE VELOCITY AT TERMINAL STANCE

| Participants | Average (deg/s) | SD (deg/s) |
|---|---|---|
| Unimpaired participants | 509.36 | 18.91 |
| Participants with disability | 342.06 | 86.52 |

TABLE II. ANGLE AT TERMINAL SWING

| Participants | Average (deg) | SD (deg) |
|---|---|---|
| Unimpaired participants | -17.76 | 8.02 |
| Participants with disability | -7.45 | 8.02 |

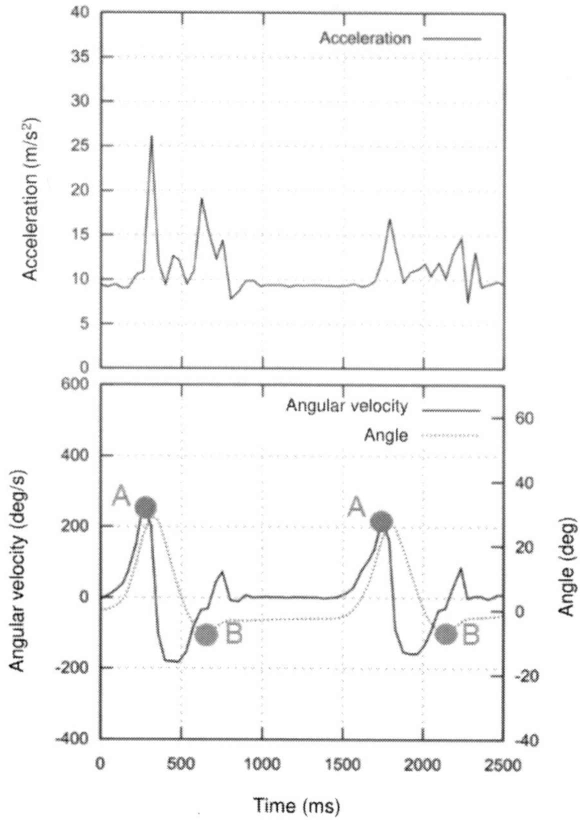


Figure 3. Examples of the angular velocity, angle, and acceleration for a participant with a walking disability.

## III. Walking Assistance Mechanisms in Passive Foot Prostheses

There are two types of walking assistance mechanisms in passive foot prostheses. First, the SACH foot [8], shown in Figure 4 [4], was designed to provide shock absorption and ankle action characteristics close to those of a normal ankle without the use of an articulated ankle joint. The SACH foot's action is achieved by the use of two functional elements: a properly shaped wedge of cushioning material built into the heel, and an internal structural core or keel shaped at the ball of the foot to provide a rocker action. A primitive version was first developed toward the end of the 1800s.

Second, the ESAR foot [5], shown in Figure 5, has a kind of leaf spring at its heel and an adequate roll-over shape like that of the foot, which generates the kicking power when raising the heel and increases the dissipated energy during the step-to-step transition in a walking gait. Wezenberg et al. reported that the ESAR foot was more effective than the SACH foot in reducing metabolic energy while walking [9], and Houdijk reported that it evolved the step length symmetry [10].

The cushioning material of the SACH foot or the leaf spring of the ESAR foot compensates for a lack of foot muscle to kick off the floor when raising the heel. This suggests that a shoe with a mechanism to raise the heel would enable elderly people with low muscle power to walk more easily.

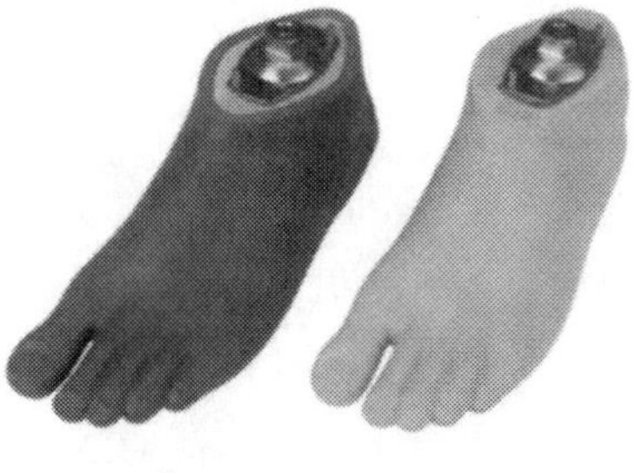

Figure 4. Examples of the SACH Foot (1D10, Ottobock, Germany).

Figure 5. Examples of the ESAR Foot (Vari-Flex, Össur, Iceland).

## IV. Spring-Assist Unit for Walking Disabilities

### A. Structure of heel-raising unit

With the goal of developing a heel-raising unit, we first developed a shoe, shown in Figure 6, to assist individuals with walking disabilities. This shoe had a coil spring and a leaf spring that enabled the wearer to raise the heel more easily. The coil spring force was 15 kg, and the shoe had a roller to prevent the toe from accidentally tripping.

In experiments with this shoe, we noticed a correlation between body weight and the most effective spring power, which would affect the walking posture. Moreover, every participant felt that the timing of generating the spring reaction force was too early for smooth walking.

Figure 6. Heel-assist shoe prototype.

We thus focused on clarifying the correlation between body weight and the optimal spring stiffness to understand its effect on walking posture. Specifically, we developed the heel-raising unit shown in Figure 7. Its mechanism was very simple, as it comprised only a conical coil spring and a V-shaped attachment cover. We adopted the conical spring so that it could be thinner. When stepped on, it provided a spring power of 3, 5, 9, or 11 kg. The attachment cover was made of thin stainless steel.

We measured electromyography (EMG) and posture data from healthy students and elderly participants by using an iEMG [11] and a Kinect [12]. The results demonstrated a linear correlation between body weight and the optimal spring stiffness, and the spring-assist unit did not affect the walking posture.

These studies were reported in the International Journal on Advances in Life Sciences in 2020 [6]. However, the device shown in Figure 7 did not have a mechanism to release the inner spring when the heel started to rise just after touching the floor in the initial contact phase.

Figure 7. Two views of the simple heel-raising unit.

We think that the inner spring should optimally release as soon as the heel rises, and this paper focuses on clarifying the effect of optimizing the spring-release timing. We thus developed the evolved heel-raising unit shown in Figure 8. This model was fabricated with a metal 3D printer. Even though it was made of titanium to be light, its weight was 210 g. The evolved unit comprises a top part A, bottom part B, inner spring, and release timing control part. The top part is placed in the heel part of a shoe. The release timing control part comprises a slider (C) U shaped rod (D), and L shaped rod (E) as shown in Figure 9.

Figure 10 illustrates the configurations for locking and releasing the inner spring. Before touching the floor, the inner spring is in the released state, as shown in Figure 10(a). When the wearer touches the heel to the floor, the top part A goes down, part C slides up, part D rotates counterclockwise, and part E rotates clockwise. Finally, as shown in Figure 10(b), the corner of part E latches with a notch in the top part A.

Next, when the wearer raises the heel from the floor, part C slides down, part E rotates clockwise, and part E rotates counterclockwise. Finally, the corner of part E releases from the notch, as shown in Figure 10(a), and the top part A raises the heel.

*B. Experimental setting*

Next, we built the evolved heel-raising unit into the heel parts of both left and right shoes, as shown in Figure 11.

To compare the difference in the assistance effect between the simple heel-raising unit and the evolved unit, we measured iEMG and motion data from healthy participants using both of them. Each participant walked with and without springs having power of 3, 5, 7, 9, and 11 kg, in the space shown in Figure 12. Wireless EMG sensors were attached to the gastrocnemius of the right leg to gather the iEMG data. The motion data was measured at the head and mid-hip to analyze the effect on the walking posture, which was measured with a Kinect.

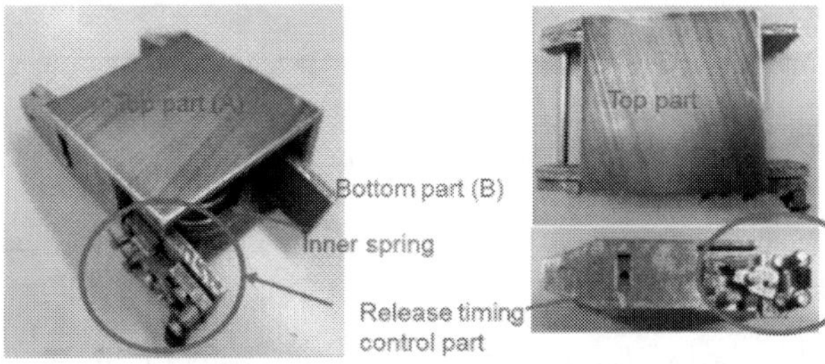


(a) Bird's-eye view (b) Top and side views

Figure 8. Various views of the evolved heel-raising unit.

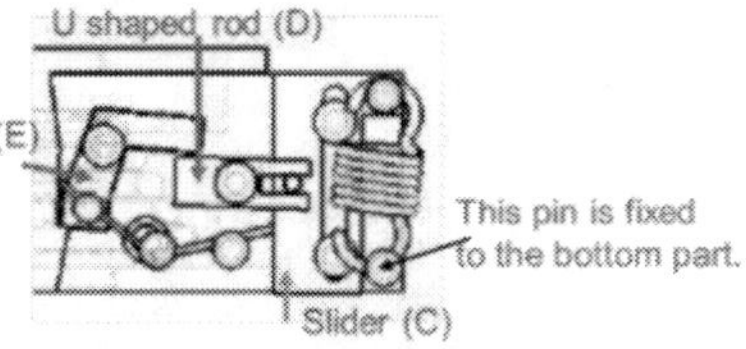


Figure 9. Structure of the release timing control part

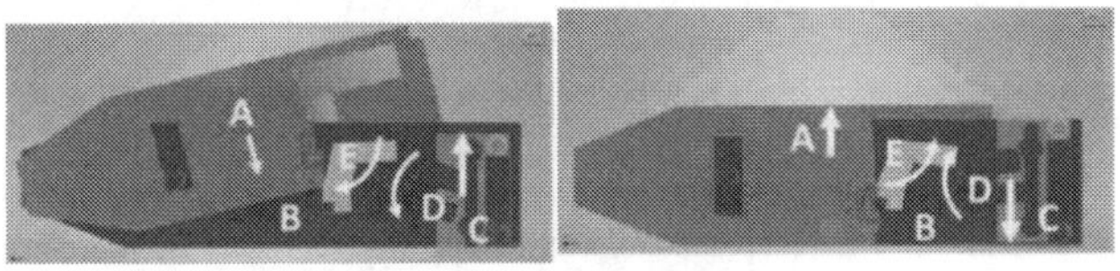

(a) Released state (b) Latched state

Figure 10. Configurations for locking and releasing the inner spring.

(a) Spring released state

(b) Spring compressed state

Figure 11. Pair of shoes incorporating the evolved heel-raising unit.

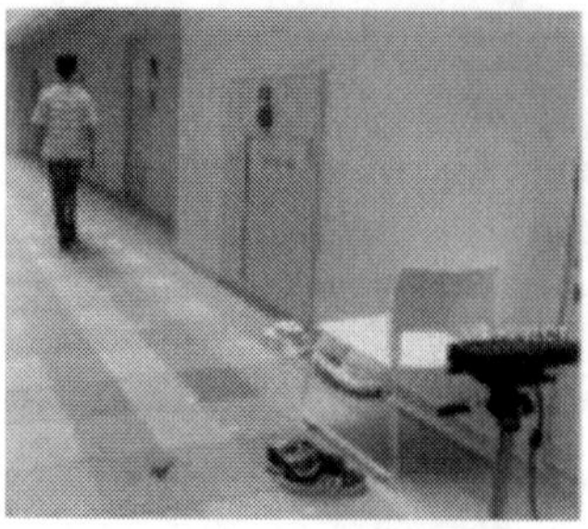

Figure 11. Measurement space.

### C. Evaluation of assistance effect

Figure 13 shows the iEMG values for each participant with each spring stiffness, and Figure 14 shows the improvement rates (IRs) due to the heel-raising unit. Here, the IR was calculated as follows:

$$IR = 100 \times (B - A)/B;$$

*A: iEMG at most effective spring stiffness;*
*B: iEMG without heel-raising unit.*

The IRs were different for each participant, but those with the evolved unit were clearly larger than those with the simple unit. Note that the measured iEMG values for participant C were much larger than those of the other participants; however, the reason for this remains unclear.

Figure 15 shows the spring stiffness at the lowest iEMG value with respect to the participant's body weight. In contrast to the case of the simple unit, the results for the evolved unit did not show a linear correlation between body weight and the optimal spring stiffness. However, because the number of participants was not enough for appropriate evaluation, we will need to measure such data for more participants.

Finally, Figure 16 shows the average peak-to-peak ranges (A-P2P) for the participants' top/bottom (T/B) and left/right (L/R) motions at the head and mid-hip. Although there were differences among the participants and differences between not using the heel-raising unit, using the simple unit and the evolved unit regarding the order of A-P2P, the differences appeared random and showed no obvious trends.

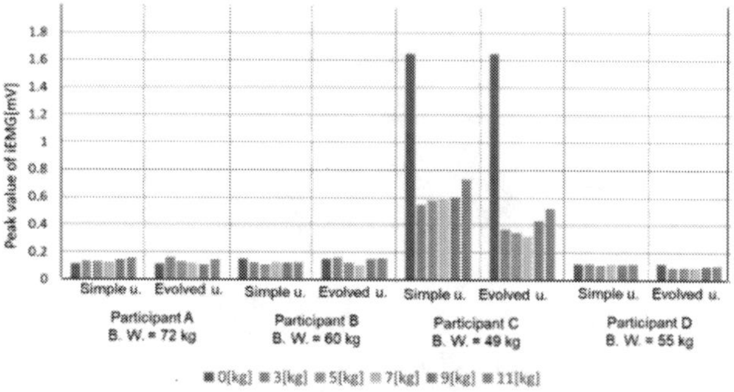


Figure 13. Measured iEMG vs. spring stiffness for each participant.

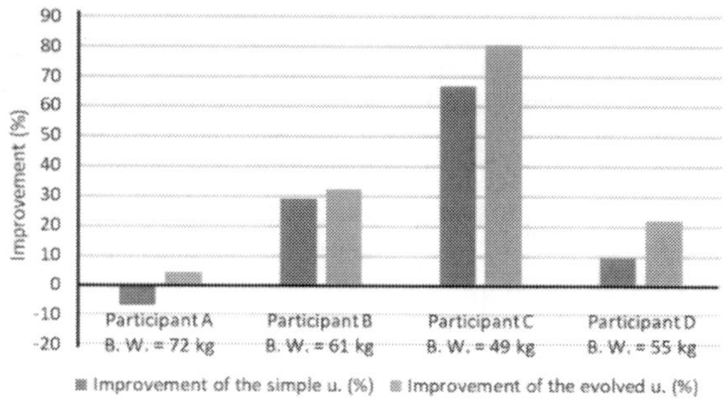


Figure 14. Improvement rate due to heel-raising unit.

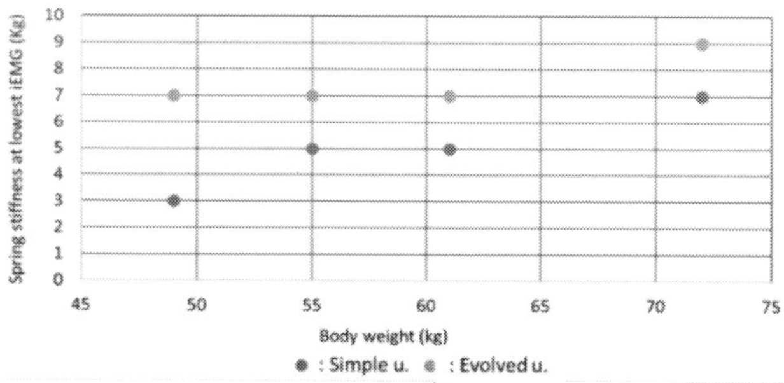


Figure 15. Spring power at lowest iEMG vs. participant body weight.

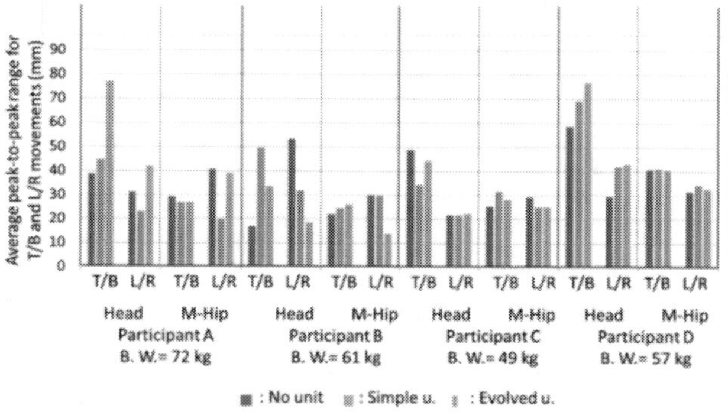


Figure 16. Average peak-to-peak range for T/B and L/R motion over two steps [mm].

The results also demonstrate that the evolved unit had little effect on the walking posture as much as the simple unit did

## V. CONCLUSIONS

It is difficult for elderly people to raise their heels because of their low muscle power. As a result, most shuffle their feet when walking and sometimes stumble or trip over something and fall down. We have thus spent nearly 10 years studying how to prevent such falls and developing a heel-raising unit to compensate for low muscle power. In this work, we developed an evolved heel-raising unit, in which a spring in the unit is released when the foot shifts from the whole shoe contacting the floor to the heel rising. Experimental results with the evolved unit demonstrated that its assistance effect was notably better than that of our previous unit, whose structure did not control the spring-release timing. Moreover, the evolved unit had little effect on the walking posture as much as the previous one had. This experiment included a low number of participants who were all healthy students. We will thus need to measure data from many elderly people to correctly evaluate the evolved unit's assistance effect. Another problem is that the prototype model in these experiments was heavy and sometimes broke or did not work well, since the model was fabricated with a metal 3D printer. A consumer model will need to be lighter and more reliable, and to have a lower cost. After developing lighter and more reliable model, we would like to measure data from many elderly people to reconfirm assistance effects of the evolved heel-raising unit.

## ACKNOWLEDGEMENTS

We thank the students in the Suzuki Lab for their help in conducting this research. We also thank Mr. Shizuo Yazawa for designing the evolved heel-raising unit. This work was supported by JSPS KAKENHI Grant Number 19K11326.

## REFERENCES

[1] World Population Ageing 2020, United Nations, [Online] https://www.un.org/development/desa/pd/sites/www.un.org.development.desa.pd/files/undesa_pd-2020_world_population_ageing_highlights.pdf [retrieved: May, 2022].

[2] W. P. Berg, H. M. Alessio, E. M. Mills, and C. Tong, "Circumstances and consequences of falls in independent community-dwelling older adults", Age Ageing, Vol. 26, pp. 261-268, 1997.

[3] Y. Murata, S. Yoshida, T. Niinuma, and K. Yoshida, "Comparative Analysis of Walking Gait Cycle between Healthy People and Walking Disabilities to Prevent Tripping Using Wearable Device and KINECT," IARIA, International Journal on Advances in Life Sciences, vol 9 no 3 & 4, pp. 186-197, 2017.

[4] Ottobock, 1D10 Dynamic foot, [Online] https://professionals.ottobock.com.au/Products/Prosthetics/Prosthetics-Lower-Limb/Feet/1D10-Dynamic-foot/p/1D10, [retrieved: May, 2022].

[5] Össur, Vari-Flex®, [Online] https://www.ossur.ca/prosthetic-solutions/products/dynamic-solutions/vari-flex [retrieved: May, 2022].

[6] H. Baba, A. Suzuki, and Y. Murata, "Spring Assist Unit for Individuals with Walking Disabilities," IARIA, International Journal on Advances in Life Sciences, vol 12 no 3 & 4, pp. 102-113, 2020.

[7] W. Tao, T. Liu, R. Zheng, and H. Feng, "Gait Analysis Using Wearable Sensors," Sensors, 12(2), pp. 2255-2283, 2012.

[8] A. Staros, "The SACH (Solid-Ankle Cushion-Heel)," Orthopedic & Prosthetic Appliance Journal, pp. 23-31, 1957.

[9] D. Wezenberg, A. G. Cutti, and H. Houdijk, "Differentiation between solid-ankle cushioned heel and energy storage and return prosthetic foot based on step-to-step transition cost," The Journal of Rehabilitation Research and Development, Volume 51, Number 10, pp. 1579-1590, 2014.

[10] H. Houdijk, D. Wezenberg, L. Hak, and A. G. Cutti, "Energy storing and return prosthetic feet improve step length symmetry while preserving margins of stability in persons with transtibial amputation," Journal of NeuroEngineering and Rehabilitation 2018, 15(Suppl 1):76, pp. 41-48, 2018.

[11] Wireless EMG logger, Logical Product Corporation, [Online] http://www.lp-d.co.jp/EMGSensor.html, [in Japanese, retrieved: May, 2022].

[12] Meet Kinect for Windows, [Online] https://developer.microsoft.com/en-us/windows/kinect, [retrieved: May, 2022].

# Measuring Perceptions of Openness in Health Information Technology Platforms

## Results from Pilot Testing Proposed Survey Framework

Kristian Malm-Nicolaisen
Norwegian Centre for E-health Research
UIT – The Arctic University of Norway
Tromsø, Norway
E-mail: Kristian.nicolaisen@ehealthresearch.no

Rune Pedersen
Norwegian Centre for E-health Research
UIT – The Arctic University of Norway
Tromsø, Norway
E-mail: Rune.pedersen@ehealthresearch.no

Asbjørn Johansen Fagerlund
Norwegian Centre for E-health Research
Tromsø, Norway
E-mail: Asbjorn.johansen.fagerlund@ehealthresearch.no

*Abstract*— **The paper focuses on the concept of Open Platforms as an emerging Health Information Technology and investigates how subjective perceptions of platform openness can be investigated across different socio-technical dimensions by piloting a novel survey framework among a small sample of domain experts. Using six main topics of inquiry, the framework was able to distinguish between type and degree of platform openness across different implementations of Health Information Technology platforms. The results indicate high aptness and relevance of the included topics. The paper proposes a step towards operationalizing the concept of Open Platforms within the e-health domain, with a special focus on patient records and interoperable systems.**

***Keywords-Health Information Technology; Open Platform; Platform Openness; Interoperability; Data Standardization.***

### I. INTRODUCTION

The concept of *Open Platforms* has emerged in the research and applied fields of Health Information Technology (HIT) domain as an elusive sub-category of traditional digital platforms. However, an authoritative definition of Open Platforms is still lacking [1]. This is a hindrance for healthcare organizations and implementers, that need to carefully consider the type and degree of openness their case requires. Furthermore, socio-technical research on real-world HIT settings requires the means to identify systems that more or less conform to the principles of Open Platforms. Therefore, a method for measuring openness over different central socio-technical platform principles is needed.

In this paper, we propose a definition of an Open Platform that we break down into six topics, which can be operationalized to measure openness in HIT platforms. The aim is to propose an early step towards establishing a measurement that can be used in real-world settings to evaluate a platform's degree of conformity towards generally accepted principles for Open Platforms. The paper describes a set of principles and a proposal on how to operationalize them in a survey format. Finally, we present the results from an initial pilot test among a small sample of expert respondents.

We have not identified any equivalent instrument or measurement tool that can measure the extent of platform openness in a real-world context, and previous studies have stated that an agreed-upon method for how to dimensionalize and measure platform openness is lacking [2]. The research objective in the present study is as follows: Conduct a pilot survey of the drafted measuring instrument among a small sample of domain experts, investigating i) whether the instrument is able to distinguish between different levels of openness in HIT platforms, and ii) if the Open Platform principles are perceived as important by the respondents.

The remainder of the paper is organized as follows. First, we introduce the relevant background and the conceptualization of Open Platforms within the e-health domain. In Section 3, we describe the method employed in the present study. Section 4 presents the results of the survey pilot test. Finally, we discuss the findings, limitations, and contribution of the paper.

### II. BACKGROUND

The digitalization of healthcare and the introduction of HIT systems has transformed the organization, provision, and planning of medical and care services. This transformation has affected organizations, healthcare professionals, patients, and decision makers alike. For clinicians, the Electronic Health Record (EHR) represents one of the most vital HIT tools for providing patients with treatment and care [3], and has ostensible benefits related to improved quality of care, patient safety and increase in productivity [4][5]. However, interoperability and data communication remains a challenge [6][7]. One reason for this is a failure to adhere to a common data representation; data from many sources must conform to a common representation that faithfully specifies certain aspects, such as the context of data, the structure of the information, and its precise meaning [8][9]. Although the usefulness and importance of standards for building shared health information networks has been proved, adoption and standardization processes have shown to be slow [10]. In addition, a lack of regulatory requirements, technology governance and supportive legislation has contributed to slowing down the progression [8].

In a recent scoping review, a combination of codes, terminologies, reference models, Clinical Information Models (CIMs) and standards were identified as the most common methods for structuring patient data [10]. Traditionally HIT generally, and EHRs specifically, are built using primarily proprietary technologies, with technical standards and CIMs developed, controlled and maintained by the vendor [1]. From a business perspective, the idea behind these monolithic mega-suite systems has been to offer most, if not all, required software functionality and features in a single solution [11]. While this approach can yield great stability, it inevitably leads to 'vendor lock-in', a situation where the vendor controls the data, and information interoperability and agility are impeded [12]. Consequently, organizations become heavily reliant on single vendors and 'closed HIT systems'. As a result, healthcare organizations struggle with a fragmented system portfolio with limited possibilities for integration, numerous legacy systems, siloed data repositories that challenge (semantic) data exchange and interoperability, and costly and complex data migrations when applications are switched [8][13]. Arguably, this is a consequence of a largely application-centric focus instead of a data-centric one [14]. In addition, monolithic suite HIT tends to assume a generative form, requiring extensive adaption and customization to fit local context. This is illustrated by the challenges, and subsequent failures, of implementation efforts of large-scale suite systems in local contexts which they were not originally designed for [15].

### A. *Defining and conceptualizing Open Platforms*

Increasingly, the market trends indicate a significant shift from the traditional single-vendor mega-suite scenario towards Open Platform-based multi-vendor system ecologies based on international and open technologies [1]. However, research concerned with the concept of platform openness within a HIT perspective is limited, with most studies assuming a domain-agnostic perspective [16]. Arguably, the inherent socio-technical complexity and multifaceted interdependencies between different stakeholders (e.g., clinicians, patients, vendors, IT staff) within the HIT and medical field, requires conceptualizations and knowledge based on the specific domain in which the platform is embedded [17].

While Open Platforms have several aspects in common with *traditional digital platforms,* as defined by Rolland et al. (2018), these definitions lack a dimension of openness which we argue distinguishes an Open Platform from traditional digital platforms. A key principle for an Open Platform is that it provides an infrastructure and core services that are extensively based on openly published standards and used in common or reused across implementations [18]. Another important distinction is that between an Open Platform and an *open-source platform*; while we can regard all open-source platforms as Open Platforms, not all Open Platforms are *open source*. The main difference is that an open-source platform makes available the source code, while an Open Platform provides the interfaces and data modelling rules but may still contain proprietary components and technologies (e.g., source code). Hence, a necessary first step is therefore to provide an objective and comprehensive definition of Open Platforms. We formulate the following working definition based on a synthesis and further development, including a broader socio-technical perspective:

*An Open Platform provides a vendor- and technology-neutral digital infrastructure and associated services that are based on open, published standards that, in principle, everyone can use to access the platform. The platform makes it possible to link applications and services from many different suppliers without discrimination, such that there is a many-to-many substitutability between applications and services, and support for data portability and sharing in defined, standardized formats using common APIs. Through a governed and modularized architecture, the platform provides flexibility by facilitating features for a heterogenous user group, supporting a multitude of use cases, and promoting innovation [13][19].*

Based on this definition, an Open Platform architecture potentially limits the challenges related to non-standard interfaces and proprietary data formats traditionally used in HITs. Building the platform on open standards, interfaces and technologies through a modularized and well-governed architecture also allows for far greater flexibility for the organization to adapt and compose their system ecology as requirements change, which the interconnected network of monolithic systems lack [20][21]. In a previous multicenter case study of hospitals that recently had implemented large-scale clinical HIT platforms, we identified the flexibility to adapt the systems to individual or group level needs as significant contributing factor in reducing user resistance and increasing perceived system usability, illustrating the importance of the flexibility characteristic of Open Platforms [22].

Building on existing literature and the definition provided above, we defined six distinct topics that should be considered when assessing the similarity between a real-world HIT case and the concept of open platforms [23]. The topics are constructed to cover multiple dimensions of the platform's architecture and services, and we use the notion of both technical and organizational aspects to expose the necessary characteristics. Each of the six topics are supplemented with a principle designed to envision the maximum extent of possible openness in a platform.

#### *1) Vendor neutrality*

Open Platforms are designed to reduce reliability on single systems and vendors by utilizing openly published standards. This contrasts with traditional HITs where proprietary technologies have been the norm. Any application that adheres to the open standards can operate, produce, and consume data on the platform, eliminating challenges with vendor lock-in and overdependence to closed systems [13][24]. This promotes continuous and agile change and growth in line with emergent user needs.

Principle: To ensure broad compatibility and interchangeability between products and technologies, the platform is based on open, available, and published technologies and standards and does not depend on individual suppliers and vendors.

*2) Flexibility of system portfolio*

By design, an Open Platform facilitates flexibility and adaptability over time [25]. This is important as the platform represents a non-static HIT expected to change. In addition, the modular architecture allows for applications to be exchanged in line with newly arising needs, e.g., the organization or users have greater access and flexibility to integrate with vendors and applications that are a better fit [11]. Using well-defined and published interfaces also allows for use in ways other than initially intended or implemented [26]. The inherent tension between flexibility and increased standardization needs special attention; standardization must be coherent and comprehensive enough to allow reuse, semantic interoperability, and common points of integration, while still allowing for differences in actors' needs, emerging requirements, and variances in workflow.

Principle: The platform allows flexibility that allows users to adopt and use applications differently and independently, based on individual or group level needs. New applications can be integrated with existing systems and can share, consume, and produce data.

*3) Clinical information models*

Common CIMs and data models are a core component of an Open Platform. CIMs are used to provide an unambiguous description of the clinical content, to represent it across applications on the platform, and support persistence and querying of structured data [27]. Hence, CIMs should provide the data in a sharable, open, and computable format. However, efficient use of CIMs depends on sufficient governance and development models; the same clinical concepts may be modelled in different ways, each of which correct [28]. User involvement in design is therefore essential, but not trouble-free; collaborative user development increases the risk for extensive negotiation processes and show of power-relations between actors, often with unpredictable outcomes [29]. This underpins the need for proper governance models for management and control of the technologies [10]. For shared HITs, such as an Open Platform, the governance should be performed centrally (e.g., by government bodies).

Principle: Applications on the platform share open information models in a way that preserves the semantics in the data when it is moved between applications.

*4) Open data*

In accordance with good patient information governance practice, the Open Platform should present all data it contains in a usable and open format. While there are different methods for operationalizing this (e.g., either by using the open format natively in data storage or through mapping and transformation from other open or proprietary formats), the purpose is to provide patient information in a form that supports interoperability, data portability and reuse [8][13].

Principle: The platform supports open data, and data is exposed as needed (subject to good information governance practice) in an open, shareable, computable format in near to real-time.

*5) Technical openness in the platform*

APIs are the technical interfaces used by applications to connect to the platform and access and interact with the data. In Open Platform the full specifications of the APIs should be published and freely available. This includes definitions of the types of data, features and functions that are provided through the APIs. Several mature open APIs exist, and ideally the Open Platform should support and provide multiple interfaces [30].

Principle: To support innovation and flexibility, open APIs that provide access to the platform and data are used.

*6) Organizational openness in the platform*

Besides technical aspects, there are organizational principles that must be fulfilled for the platform to be truly open. As healthcare is a highly regulated sector, it is possible that local, national, or international laws and regulations could prohibit total openness. We therefore consider this to be a context-sensitive principle, and implementers and providers should thoroughly assess the desired, necessary, and permitted degree of openness.

Principle: Organizational, financial, and legal frameworks does not prevent third-party suppliers and the integration of additional functionality or applications.

## III. Methods

### A. Survey development and design

A survey based on the six topics presented in the previous section was developed in an iterate process between the authors. Complementary statements have been defined for each main topic for further probing and exposure of the specific topic. The statements were operationalized into 5-point bipolar Likert items that could be administered in a web-based survey (Table I) [31]. For five of the six topics, multiple Likert items were formulated in order to allow for more extensive probing. A web-based questionnaire was developed in Questback (Questback AS, Oslo, Norway). For each item, respondents gave their response over two dimensions: 1) Extent of presence in the platform: To which extent do you agree that the statement is true for your hospital's HIT platform, and 2) Overall importance: To which extent do you agree that the statement is important for an open HIT platform.

The method employed in this study is designed to measure openness as a subjective construct in the eyes of domain experts [32]. First, we wanted to measure the respondent's perception of the extent to which their HIT platform corresponded to the Open Platform principles. Secondly, we were interested in testing the respondents' perception of the Likert items' relevance for Open Platforms in general. The resulting questionnaire had 16 items, that each included both a presence and an importance rating. The survey was piloted among two domain experts and adjusted based on feedback prior to finalization. We assess the survey by evaluating whether the measurement can distinguish between varying degrees of adherence to the Open Platform principles between systems, and perceived face validity of the survey items. We explore this by considering two variables: i) presence of either floor or ceiling effects, indicating that items might be able to discriminate, and ii) items score on perceived importance to assess face validity.

TABLE I. MAIN TOPICS AND CORRESPONDING ITEMS USED IN SURVEY, INCLUDING KEY REFERENCES

| Topic | Code | Item |
|---|---|---|
| Vendor neutrality [33][34] | V1 | The platform builds upon open and non-proprietary technologies |
| | V2 | The platform builds upon open standards, such as HL7 FHIR, IHE-XDS and openEHR |
| | V3 | The platform is not dependent on a single vendor |
| Flexibility in system portfolio [35][36] | F1 | The platform allows flexibility for individual users to use applications based on their needs and preferences |
| | F2 | The platform allows flexibility for sub-divisions of the organization (i.e., departments, clinics etc.) to use applications based on their needs and preferences |
| | F3 | The platform allows new applications to be integrated with the existing systems on the platform and share/consume/produce data |
| Clinical information models [1][28][37] | I1 | The specifications for information model and terminologies are openly available |
| | I2 | The applications on the platform share information models such that the semantics in the data is preserved when moved between applications |
| Open data [38][39] | OD | Data is separated from application, in the sense that data is available in a readable, open and shareable format regardless of vendor or application |
| Technical openness in the platform [40][41][42] | T1 | The types of data, features and functions that are offered through open APIs are defined |
| | T2 | API specifications are published and available |
| Organizational openness in the platform [2][17][43] | O1 | It is clearly defined who has access and what is required to further develop core functionality on the platform |
| | O2 | It is clearly defined who has access and what is required to develop additional functionality (e.g., decision support, apps, analytical tools) on the platform |
| | O3 | It is clearly defined who has access to use data from the platform |
| | O4 | Requirements for compliance with technical standards or payment of license fees - are reasonable and non-discriminatory, that is, they are used uniformly for all potential platform participants |
| | O5 | The hospital IT services have freedom to modify, query and map the information scheme of the platform with no added licensing cost |

## B. *Recruitment to survey and data analysis*

We identified nine hospital organizations across five different European countries relevant for the survey. The sample group was identified based on prior work by the researchers [1] and consisted of hospitals using both suite IT systems and systems based on open specifications, comprising systems from five different vendors. All hospitals delivered specialized care and were organized as public providers in socialized healthcare settings.

Participation invitations were sent to high-level managers in the health informatics domain of each hospital organization. From each of the hospitals, we invited one respondent to participate, in total nine individuals were invited. Electronic questionnaires were distributed to the respondents by email, starting on the 16th of June, and closing on 17th of August 2020. Data was exported from Questback into a spreadsheet software (Microsoft Excel for Mac OS, Microsoft 365) and structured for analysis. Analyses were conducted in SPSS (IBM SPSS for Mac OS, version 26.0.0.1). To explore the range and the central tendency in the responses, minimum, maximum, and median scores were calculated [44].

# IV. RESULTS

Out of the nine respondents that were invited to participate, four completed the survey within the timeframe, yielding a response rate of 44%. The respondents of the completed surveys were from three different countries (Norway, Denmark, and the United Kingdom), using three different HIT systems. One organization refused participation based on perceived lack of study relevance. Two had opened the survey but did not complete it. Two did not open the survey. See Table II for results, interpretation, and value distribution.

Overall, calculated median scores from all respondents show a degree of presence between Disagree and Agree for the items relating to vendor neutrality (V1-V3), while the same items have received a rating of Agree to Strongly Agree for importance. All items relating to flexibility in system portfolio (F1-F3) are rated as Agree in terms of presence, and Strongly Agree for importance. Items corresponding to the topic on CIMs (I1-I2) are rated as Agree for presence and Strongly Agree for importance. The one item on Open Data (OD1) is rated Agree for presence, while it was rated Strongly Agree for importance. Both items relating to Technical Openness (T1-T2) received a median value translating to Agree for the presence dimension, and Strongly Agree for importance. Value scores on the item in the category Organizational Openness (O1-O5) varied from Neutral to Agree, with the item describing access to core platform functionality receiving the lowest score. The same items received values translating to Agree and Strongly Agree on importance.

The minimum and maximum results for the presence dimension indicate neither floor effect nor ceiling effect. Further, the results indicate that the principles were present only to a limited degree in existing HIT platforms. Overall, for the importance dimension, all items received median scores translating to Agree (18.75%) and Strongly Agree (81.25%) indicating strong relevance and aptness.

TABLE II. SURVEY RESULTS OVER BOTH DIMENSIONS, SHOWING MEDIAN (MDN) AND MIN-MAX VALUES FOR EACH LIKERT ITEM. LIKERT SCALE INTERPRETATION AND DISTRIBUTION OF VALUES: 1 STRONGLY DISAGREE (1.0-1,49); 2 DISAGREE (1.5-2.49); 3 NEUTRAL (2.5-3.49); 4 AGREE (3.5-4.49); 5 STRONGLY AGREE (4.5-5.0).

| Topic | Code | Presence | | | Importance | | |
|---|---|---|---|---|---|---|---|
| | | Mdn | Min | Max | Mdn | Min | Max |
| Vendor neutrality | V1 | 3 | 2 | 4 | 5 | 4 | 5 |
| | V2 | 4 | 3 | 5 | 5 | 4 | 5 |
| | V3 | 2 | 1 | 4 | 3.5 | 3 | 5 |
| Flexibility in system portfolio | F1 | 3.5 | 2 | 4 | 4.5 | 2 | 5 |
| | F2 | 3.5 | 3 | 4 | 5 | 4 | 5 |
| | F3 | 3.5 | 2 | 4 | 4.5 | 4 | 5 |
| Clinical information models | I1 | 4 | 2 | 5 | 5 | 4 | 5 |
| | I2 | 4 | 2 | 4 | 5 | 4 | 5 |
| Open data | OD1 | 3 | 1 | 5 | 5 | 4 | 5 |
| Technical openness in the platform | T1 | 4 | 2 | 4 | 4.5 | 4 | 5 |
| | T2 | 4 | 2 | 4 | 4.5 | 4 | 5 |
| Organizational openness in the platform | O1 | 2.5 | 2 | 4 | 4 | 4 | 5 |
| | O2 | 3 | 1 | 4 | 4 | 4 | 5 |
| | O3 | 3.5 | 3 | 5 | 5 | 4 | 5 |
| | O4 | 4 | 3 | 5 | 4.5 | 4 | 5 |
| | O5 | 4 | 1 | 5 | 5 | 4 | 5 |

## V. CONCLUSION AND FUTURE WORK

In this paper, we were able to pilot the measuring tool among a small number of domain experts. Even though we consider them highly informed respondents, it is likely that further drivers, challenges, and topics should be considered. While the study was designed to only include a small number of respondents, the survey achieved a response rate comparable to other internet based surveys aimed towards professionals [45]. The respondents in this survey were a convenience sample of mid-to-high level professionals. The results suggested that the measuring tool was able to distinguish between different degrees of open platform presence in the surveyed HIT implementations. The range between minimum and maximum scores indicated that on most items, the respondents had not answered uniformly. Furthermore, the results indicated that the respondents perceived the items as important. The median item scores indicate that the included platforms are positioned far from the end of an openness spectrum. This corresponds well with the actual platform market landscape we can observe; Open Platforms as a HIT approach are still in the early stages, with some immature technological traits and lack of supporting regulations and governance models.

The present study has some clear but intended limitations and should be considered an early proposal on how to move towards operationalization and application of concepts that originates in a highly theoretical field of research. Real-world implementation and use of Open Platforms are complicated at several levels. Our framework for measuring perceptions of openness deliberately excludes topics relating to regulatory aspects and law as these are highly context sensitive and vary greatly. The miniscule sample of respondents rules out the possibility for any meaningful ordinary validation procedure with measurements of internal consistency and factor analyses. However, given the consistent high importance valuation of the items by the respondents, we consider it worthwhile to further investigate the Open Platform concept within this framework. This study resides within a larger research portfolio interested in design and implementation of shared large-scale HIT platforms. Based on the work presented in this study, we have conducted a qualitative study investigating the Open Platform principles in relation to an ongoing national HIT platform project in Norway. The topics and statements used in the framework guided the development of an interview guide, and we were able to investigate how the properties aligned with a HIT project described as an Open Platform. While not yet published, the results indicates that an ambiguous and disparate understanding of the concept of Open Platforms among different stakeholders created misunderstandings and contrasting expectations, and that the proposed HIT platform shared few of the characteristics outlined in our framework, although being described as an "open platform" [23]. While the present study is limited to the perspective of health informatics experts, we widen the scope to also include healthcare professionals and HIT vendors in our related work. The final results from the qualitative study will be used to further develop the framework presented in the present paper, establishing the basis for employing a larger survey and enabling a validation procedure. A validated set of measurable principles can provide important insight and grounds for a common and unambiguous concept understanding for implementors and organizations considering an Open Platform approach in their HIT infrastructure.

## ACKNOWLEDGMENT

This study was funded by the Norwegian Centre for E-health Research.

## REFERENCES

[1] K. Malm-Nicolaisen, R. Pedersen, and A. J. Fagerlund, «Open or closed: A project proposal for investigating two different EHR platform approaches.» Context Sensitive Health Informatics: Sustainability in Dynamic Ecosystems, vol. 265, no., pp. 207-212, 2019.

[2] L. D. Thomas, E. Autio, and D. M. Gann, «Architectural leverage: Putting platforms in context.» Academy of management perspectives, vol. 28, no. 2, pp. 198-219, 2014.

[3] R. F. Gillum, «From papyrus to the electronic tablet: a brief history of the clinical medical record with lessons for the digital age.» The American journal of medicine, vol. 126, no. 10, pp. 853-857, 2013.

[4] R. Hillestad, J. Bigelow, A. Bower, F. Girosi, R. Meili, R. Scoville, et al., «Can electronic medical record systems transform health care? Potential health benefits, savings, and costs.» Health affairs, vol. 24, no. 5, pp. 1103-1117, 2005.

[5] B. Chaudhry, J. Wang, S. Wu, M. Maglione, W. Mojica, E. Roth, et al., «Systematic review: impact of health information technology on quality, efficiency, and costs of medical care.» Annals of internal medicine, vol. 144, no. 10, pp. 742-752, 2006.

[6] J. M. McGinnis, D. Aisner, and L. Olsen, «The learning healthcare system: workshop summary.». 2007: National Academies Press.

[7] C. P. Friedman, A. K. Wong, and D. Blumenthal, «Achieving a nationwide learning health system.» Science Translational Medicine, vol. 2, no. 57, pp. 57cm29, 2010.

[8] L. Marco-Ruiz, K. Malm-Nicolaisen, R. Pedersen, A. Makhlysheva, and P. Bakkevoll, «Ontology-based terminologies for healthcare.» Nasjonalt senter for e-helseforskning, no., pp. 78, 2017.

[9] E. V. Bernstam, J. L. Warner, J. C. Krauss, E. Ambinder, W. S. Rubinstein, G. Komatsoulis, et al., «Quantitating and assessing interoperability between electronic health records.» Journal of the American Medical Informatics Association, no., 2022.

[10] K. Malm-Nicolaisen, L. M. Ruiz, E. R. Evenstad, and R. Pedersen. «Efforts on Using Standards for Defining the Structuring of Electronic Health Record Data: A Scoping Review.» Proceedings of the 17th Scandinavian Conference on Health Informatics, November 12-13, 2019.: Linköping University Electronic Press. pp. 108-115.

[11] A. Williams. *Best of Breed vs. Monolithic Systems: Finding the Best Software Solutions Philosophy*. 2018 [cited 2022 03.15]; Available from: https://www.mapcom.com/blog/best-of-breed-vs-monolithic-systems-finding-the-best-software-solutions-philosophy/.

[12] A. Byrne. *Can the road to open platform run alongside closed systems?* 2017 [cited 2022 03.06]; Available from: https://www.digitalhealth.net/2017/11/can-the-road-to-open-platform-run-alongside-proprietary-systems/.

[13] Apperta Foundation CIC. *Defining an Open Platform*. 2017 [cited 2022 02.15]; Available from: https://apperta.org/assets/Apperta_Defining_an_Open_Platform.pdf.

[14] T. Gornik. *The Postmodern EHR: The Data Layer*. 2016 [cited 2022 03.05]; Available from: https://www.openhealthnews.com/story/2016-05-04/postmodern-ehr-data-layer.

[15] R. Bergstrøm. *Demanding to implement e-health solutions from the USA (Krevende å implementere helseløsninger fra USA)*. [cited 2022 02.21]; Available from: https://www.dagensmedisin.no/artikler/2019/11/25/kulturforskjeller-skaper-it-trobbel/.

[16] P. Van Gorp, M. Comuzzi, A. Jahnen, U. Kaymak, and B. Middleton, «An open platform for personal health record apps with platform-level privacy protection.» Computers in biology and medicine, vol. 51, no., pp. 14-23, 2014.

[17] M. De Reuver, C. Sørensen, and R. C. Basole, «The digital platform: a research agenda.» Journal of Information Technology, vol. 33, no. 2, pp. 124-135, 2018.

[18] K. Boudreau, «Open platform strategies and innovation: Granting access vs. devolving control.» Management science, vol. 56, no. 10, pp. 1849-1872, 2010.

[19] G. C. Bowker and S. L. Star, «Sorting things out: Classification and its consequences.». 2000: MIT press.

[20] M. Aanestad and P. Vassilakopoulou. «Innovation Readiness in Healthcare Information Infrastructures. Key Resources to Enable Collaborative Digital Innovation.» Proceedings of the 17th Scandinavian Conference on Health Informatics, 12 -13 Nov 2019.: Linköping University Press. pp. 61-66.

[21] K. H. Rolland, L. Mathiassen, and A. Rai, «Managing digital platforms in user organizations: the interactions between digital options and digital debt.» Information Systems Research, vol. 29, no. 2, pp. 419-443, 2018.

[22] K. Malm-Nicolaisen, A. J. Fagerlund, and R. Pedersen, «How does users of modern EHR perceive the usability, user resistance and productivity five years or more after implementation?» MedInfo 2021, In press, no., 2021.

[23] K. Malm-Nicolaisen, A. J. Fagerlund, and R. Pedersen, «Openness in Health Information Technologies: Conceptualizing and Leveraging Flexibility in Open Platforms.» Unpublished, no.

[24] J. Hoeksma. *Gartner says NHS over-reliant on closed systems*. Digital Health 2017 [cited 2022 04.23]; Available from: https://www.digitalhealth.net/2017/09/gartner-says-nhs-reliant-closed-systems/.

[25] T. Beale. *What is an 'open platform'?* 2014 [cited 2022 04.25]; Available from: https://wolandscat.net/2014/05/07/what-is-an-open-platform/.

[26] D. Furstenau and C. Auschra. «Open digital platforms in health care: Implementation and scaling strategies.» Thirty Seventh International Conference on Information Systems, 2016. pp. 1-12.

[27] Inidus. *What is an Open Platform?* [cited 2022 03.26]; Available from: https://inidus.com/what-is-an-open-platform/.

[28] T. A. Oniki, J. F. Coyle, C. G. Parker, and S. M. Huff, «Lessons learned in detailed clinical modeling at Intermountain Healthcare.» Journal of the American Medical Informatics Association, vol. 21, no. 6, pp. 1076-1081, 2014.

[29] B. Latour, «Science in action: How to follow scientists and engineers through society.». 1987: Harvard University Press.

[30] E. Davis. *Defining An Open Platform for Health IT*. 2016 [cited 2022 04.15]; Available from: https://www.openhealthnews.com/story/2016-03-12/defining-open-platform-health-it.

[31] R. Likert, «A technique for the measurement of attitudes.» Archives of Psychology, no., 1932.

[32] A. Benlian, D. Hilkert, and T. Hess, «How open is this Platform? The meaning and measurement of platform openness from the complementers' perspective.» Journal of Information Technology, vol. 30, no. 3, pp. 209-228, 2015.

[33] G. Ellingsen, B. Christensen, and L. Silsand, «Developing large-scale electronic patient records conforming to the openEHR architecture.» Procedia Technology, vol. 16, no., pp. 1281-1286, 2014.

[34] A. Gawer and M. A. Cusumano, «Industry platforms and ecosystem innovation.» Journal of product innovation management, vol. 31, no. 3, pp. 417-433, 2014.

[35] A. Gawer, «Bridging differing perspectives on technological platforms: Toward an integrative framework.» Research policy, vol. 43, no. 7, pp. 1239-1249, 2014.

[36] D. Tilson, C. Sorensen, and K. Lyytinen. «Change and control paradoxes in mobile infrastructure innovation: the Android and iOS mobile operating systems cases.» 45th Hawaii International Conference on System Sciences, 2012. IEEE. pp. 1324-1333.

[37] G. Ellingsen and E. Monteiro, «Seamless integration: standardisation across multiple local settings.» Computer

Supported Cooperative Work (CSCW), vol. 15, no. 5-6, pp. 443-466, 2006.

[38] G.-H. Ulriksen, R. Pedersen, and G. Ellingsen, «Infrastructuring in healthcare through the openEHR architecture.» Computer Supported Cooperative Work (CSCW), vol. 26, no. 1-2, pp. 33-69, 2017.

[39] K. Vestues and K. Rolland, «Platformizing the Organization through Decoupling and Recoupling: A longitudinal case study of a government agency.» Scandinavian Journal of Information Systems, vol. 33, no. 1, pp. 5, 2021.

[40] D. Soto Setzke, M. Böhm, and H. Krcmar. «Platform openness: A systematic literature review and avenues for future research.» 14th International Conference on Wirtschaftsinformatik, February 24-27, 2019.

[41] A. Ghazawneh and O. Henfridsson, «Balancing platform control and external contribution in third - party development: the boundary resources model.» Information systems journal, vol. 23, no. 2, pp. 173-192, 2013.

[42] K. Kapoor, A. Z. Bigdeli, Y. K. Dwivedi, A. Schroeder, A. Beltagui, and T. Baines, «A socio-technical view of platform ecosystems: Systematic review and research agenda.» Journal of Business Research, vol. 128, no., pp. 94-108, 2021.

[43] T. R. Eisenmann, G. Parker, and M. Van Alstyne, «Opening platforms: how, when and why?» Platforms, markets and innovation, vol. 6, no., pp. 131-162, 2009.

[44] H. N. Boone and D. A. Boone, «Analyzing likert data.» Journal of extension, vol. 50, no. 2, pp. 1-5, 2012.

[45] G. Yetter and K. Capaccioli, «Differences in responses to web and paper surveys among school professionals.» Behavior Research Methods, vol. 42, no. 1, pp. 266-272, 2010.

# Information Technology Self-Efficacy and Confidence Amongst Health Professions Students Enrolling in a Telehealth Educational Course

Rodrigo Mariño
Melbourne Dental School
University of Melbourne
Melbourne, Australia
e-mail: r.marino@unimelb.edu.au

Mark Merolli
Dept of Physiotherapy, School of Health Sciences,
University of Melbourne; Centre for Digital Transformation of Health
Melbourne, Australia
e-mail: merollim@unimelb.edu.au

Daniel Capurro
Melbourne School of Engineering
School of Computing and Information Systems
University of Melbourne; Centre for Digital Transformation of Health
Melbourne, Australia
e-mail: dcapurro@unimelb.edu.au

***Abstract*—A telehealth course was designed to provide entry-to-practice health professions students with a set of foundational knowledge and skills needed to succeed as a virtual health care practitioner in an age where remote care has become increasingly needed. In preparation for the large-scale testing and validation of the course, this study examined the level of technology proficiency (reported as self-assessed 'confidence') amongst the future healthcare workforce. This assessment was undertaken among students from four health sciences departments at the University of Melbourne, Australia. Students were invited to participate in the study by completing an online, anonymous, 68-item questionnaire. Relevant to this manuscript, the questionnaire included 22 items pertaining to Internet and Information Communication Technology (ICT) use, adapted from Technology Proficiency Self-Assessment Questionnaire for 21st Century Learning (TPSA C-21). Results indicated that access to technology and frequency of use of the Internet was high, with 57.7% of the students accessing the Internet, at least, every 30 minutes. When students were asked about their level of confidence with ICT, responses suggested that students were confident in their ability to perform all the task included in the TPSA C-21. Nonetheless, students felt less confident with the administrative aspects of mobile technology and in its use as a tool for their future role as health professionals. Findings indicate that students have adequate proficiency. The study identified some areas in which support, and further development may be required.**

***Keywords – students; use of ICT; technology proficiency***

## I. INTRODUCTION

Telehealth (telemedicine, virtual care, etc.) has developed and continues to advance rapidly for several reasons. These include but are not limited to progress in Information and Communication Technologies (ICT), increasing patient's expectations and preferences, and a need for flexible models of care [1][2]. Telehealth is a useful vehicle to improve equity of access and opportunity for care to patients, help solve specific health problems, and expand the possibilities of continuous training/furthering development for health professionals [1].

More recently, it has been observed how, in a short period of time, due to the COVID-19 pandemic, an expansion in the use of telehealth and ICT from all its modalities of practices, to ensure that health services are provided and made available to the population [3][4]. This situation has highlighted the urgent need and obligation to train our future health workforce to new, modalities of practice, which undoubtedly will involve some form of telehealth to operate effectively [5][6]. Health professionals of the digital age must develop proficiencies and competencies in this area and understand its opportunities and limitations of working under this "new normal".

The Faculty of Medicine, Dentistry, and Health Sciences, The University of Melbourne developed and evaluated a blended learning experience to provide entry-to-practice health professions students with a set of foundational skills and knowledge needed to succeed as a virtual health care practitioner [7]. This learning experience will equip health professionals with the core capabilities that are necessary to work in diverse multidisciplinary scenarios and address current and new demands in health care using telehealth and ICT.

The modules in this course provide access to theoretical elements of the state of the art in telehealth and generate spaces for reflection that allow health professional students to acquire the basic core knowledge to go beyond teleconsultation and remote triage. The course also allows students to identify those healthcare services and procedures that could be successfully provided through telehealth. The emphasis is not on technology, but on the ability of students to:

- communicate with various stakeholders (i.e., patients, other clinicians, care teams)
- deliver patient-centered education and care
- effectively adopt new technologies
- identify barriers and facilitators to care using this model
- work in multidisciplinary healthcare teams.

This paper is organized in five sections. The first Section provides the foundation for understanding the need for a telehealth course and the structure of such a course and presents the aims and objectives of the study. Section II describes the methodology used in the pilot test the course. Section III presents the results of the trial, including students involved in the course. Sections IV and V discuss the results of the trial and conclude on its findings, respectively. Further steps are also discussed in last section.

The specific objective of this study was to assess the level of self-assessed technology proficiency amongst the future healthcare workforce. This assessment was undertaken with students across four interprofessional health sciences Departments/Schools at the University of Melbourne, Australia: Dentistry, Physiotherapy, Social Work, and Audiology and Speech Pathology.

This evaluation was considered an important first step alongside an understanding of implementation challenges and associated issues to aid in integrating telehealth into the educational environment.

## II. METHODS

With the approval of the Human Research Ethics Committee from The University of Melbourne (Study ID: 20529), students from five health science award courses (across four departments) at the University of Melbourne were recruited to participate in this project. These courses were: Dentistry, Oral Health Therapy, Physiotherapy, Social Work, and Clinical Audiology.

Data was collected between August and September 2021. During this period students were invited to participate in the study by first completing an online anonymous, 68-item questionnaire. The questionnaire included items on socio-demographic characteristics and course level data, as well as Internet utilization information (i.e., frequency and devices used). Additionally, the instrument included 22 items asking about Internet and information communication technology (ICT) use, adapted from Technology Proficiency Self-Assessment Questionnaire for 21st Century Learning (TPSA C-21) [8]. The instrument also contained a subsequent 34 items about perceptions surrounding using telehealth [9]. Students' perceptions around telehealth were captured according to the Unified Theory of Acceptance and Use of Technology (UTAUT2). The UTAUT2 is a validated and reliable framework for examining elements of technology acceptance and has been adopted in several studies [10][11].

Socio-demographic information reported in this analysis included age, sex, and course of study.

Internet utilization information included: Frequency of online access; participants classified themselves according to frequency of visit to Internet sites, as 'At least hourly, 'At least daily, 'At least weekly', 'At least monthly', and 'Less than once a month'. The device use list included 9 alternatives: Mobile smartphone; tablet (i.e., iPad); desktop computer; laptop; smart TV; gaming console (i.e., Xbox, PlayStation); smartwatch (i.e., Apple watch, Fitbit); eBook reader (i.e., Kindle, Kobo, etc.); and smart home assistant (i.e., Google Home, Amazon Alexa).

The TPSA C-21 was assessed on a 5-point ordinal Likert scale, according to the response that best described their confidence in using the Internet and ICT as 'Strongly disagree' to 'Strongly agree'. These responses were weighted as 'Somewhat agree'= 0.5, 'Strongly agree'= 1, all other responses = 0. The weighted values were added to yield a total confidence (proficiency) score.

Sample size calculation were based on the minimum requirements to detect a change from pre- to post-intervention (i.e., participation in the telehealth course), it was estimated that a total sample size of 26 participants would be necessary to detect a mean difference of five-tenths (0.50) of the standard deviation in major outcomes between paired observations of participants (pre-test vs. post-test), at the uni-dimensional significance criterion of 0.05, and a power of 0.80 [12]. The study used convenience samples of male and female students, 18 years or older, enrolled in the aforementioned courses.

Data were analysed using IBM SPSS to statistically compare results between different socio-demographic and Internet use variables. Given the small sample size, only basic descriptive information on the distribution of selected socio-demographic and study variables was performed. Categorical and ordinal variables were analysed utilizing Chi square analysis ($\chi^2$). For continuous variables (technology proficiency results), data were analysed using one-way analysis of variance (ANOVA) to examine main effects of each of the independent variables on the dependent variable under study.

## III. RESULTS

In total, 26 students enrolled in the training course and completed the pre-assessment instrument. The majority (82.6%) were female and aged 29 years-old or younger (76.9%). Nine participants were from the Dental School (34.6%), 23.1% were from Physiotherapy, another 23.1% were Social Work students, and the remaining 19.2% were Audiology students.

All participants had access to a smartphone and a laptop. Frequency of use of the Internet was high (Table I), 57.7% of the students accessed the Internet at least, every 30 minutes. Another 23.1% access very hour. No statistically significant differences in frequency of use were found by profession, sex, or age group.

TABLE I. FREQUENCY OF ONLINE ACCESS (%)[a]

| Every 10 minutes | Every 30 minutes | Every hour | at least once a day |
|---|---|---|---|
| 42.3 | 15.4 | 23.1 | 19.2 |

a. N= 26

When students were asked about their level of confidence with technology (their self-assessed proficiency), results indicated (Table II) that they were confident in their ability to perform most of the tasks. Confidence scores ranged from 13.0 to 22.0, with an overall mean of 19.8 (s.d. 2.5). Half of the participants scored 18.0 points or more in the confidence scale. There were no significant differences by gender, age group, frequency of use, or profession.

Except for one item (item 11), participants scored highly in all the items (>55.0%). In particular, they all (100%) strongly agreed that they were able to: find webpages related to my subject matter interests (Item 1); use the computer to create a slideshow presentation (Item 7); and download and view streaming movies/video clips (Item 20). In another seven items, all students were either Strongly or Somewhat confident that they would (Items 3,6,9,18,19,21,22); and that they confidently would download podcasts and audio books; or send phots via a smartphone; or and safe and retrieve files from the cloud.

On the other hand, although the majority (> 50%) were strongly/somewhat confident in their proficiency, they were less confident in areas requiring deeper skills such as: creating a database of information (26.9%) (Item 8); integrating mobile technology in their work or creating a blog (26.9%) (Item 14). Students were also less confident in describing software programs or apps they would use in their role as healthcare professionals (19.2%) (Item 10), and to a lesser extent using social media tools as part of their role as health professional (7.6%) (Item 13).

Healthcare profession students felt less confident in two items: writing a plan with a budget to buy technology that would support me in my role as health professional (Item 11); and on how to create their own webpage (Item 2). When asked about how satisfied they were writing a plan with a budget to buy technology that would support me in my role as health professional, although the majority was either somewhat confident or slightly confident (57.7%), 15.4% was neutral and, more importantly, another 26.9% somewhat or strongly disagree with the statement.

The majority were also confident (strongly: 30.8%; or somewhat: 30.8%) that they could create their own webpage. However, 15.4% were neutral, and another 19.2% were somewhat confident and 3.8% was not confident.

TABLE II. STUDENTS' RESPONSES TO TECHNOLOGY PROFICIENCY QUESTIONNAIRE (%)[a]

I FEEL CONFIDENT THAT I COULD....

| Strongly disagree | Somewhat disagree | Neutral | Somewhat agree | Strongly agree |
|---|---|---|---|---|
| 1. Use an Internet search engine (e.g., Google) to find webpages related to my subject matter interests?[a] | | | | |
| - | - | - | - | 100.0 |
| 2. Create my own webpage? | | | | |
| 3.8 | 19.2 | 15.4 | 30.8 | 30.8 |
| 3. Find primary sources of information on the Internet that I can use in my role as a health professional | | | | |
| - | - | - | 11.5 | 88.5 |
| 4. Use a spreadsheet to create a bar graph? | | | | |
| - | - | 3.8 | 26.9 | 69.2 |
| 5. Create a newsletter with graphics | | | | |
| - | - | 11.5 | 38.5 | 50.0 |
| 6. Save documents in formats so that others can read them if they have different word processing programs | | | | |
| - | - | - | 7.7 | 92.3 |
| 7. Use the computer to create a slideshow presentation | | | | |
| - | - | - | - | 100.0 |
| 8. Create a database of information | | | | |
| - | 7.7 | 19.2 | 23.1 | 50.0 |
| 9. Use technology to collaborate with other people who are distant from where I am | | | | |
| - | - | - | 3.8 | 96.2 |
| 10. Describe 5 software programs or apps that I would use in my role as health professional | | | | |
| - | 11.5 | 7.7 | 30.8 | 50.0 |
| 11. Write a plan with a budget to buy technology that would support me in my role as health professional | | | | |
| 11.5 | 15.4 | 15.4 | 34.6 | 23.1 |
| 12. Integrate mobile technologies into my role as health professional | | | | |
| - | 3.8 | 23.1 | 19.2 | 53.8 |
| 13. Use social media tools as part of my role as health professional | | | | |
| - | 3.8 | 3.8 | 23.1 | 69.2 |
| 14. Create a wiki or blog to have peers collaborate | | | | |
| - | 11.5 | 15.4 | 34.6 | 38.5 |
| 15. Use online tools to communicate from a distance in my role as health professional | | | | |
| - | - | 3.8 | 15.4 | 80.8 |
| 16. Communicate with someone in a one-to-one environment in which the other person has their own device | | | | |
| - | - | 11.5 | 11.5 | 76.9 |
| 17. Find a way to use a smartphone in my dole as a health professional to collect people's responses | | | | |
| - | - | 7.7 | 19.2 | 73.1 |
| 18. Use mobile devices to connect to others for my professional development | | | | |
| - | - | - | 19.2 | 80.8 |
| 19. Download and listen to podcasts/audio books | | | | |
| - | - | - | 3.8 | 96.2 |
| 20. Download and view streaming movies/video clips | | | | |
| - | - | - | - | 100.0 |
| 21. Send/transfer photos or other data via a smartphone | | | | |
| - | - | - | 3.8 | 96.2 |
| 22. Save and retrieve files in a cloud-based environment | | | | |
| - | - | - | 19.2 | 80.8 |

b. N=26

## IV. Discussion

The ability to integrate 21st century technology for learning is an expectation for educators. If educators want to ensure that the future healthcare work forces are digitally trained, then self-efficacy and technological abilities are important constructs [13]. The information provided by this cohort of students would indicate that students in healthcare professional courses at the University of Melbourne are highly proficient, and able to use a wide range of technologies regularly in their daily lives.

This was a cohort of students who have been exposed to ICT in education since, at least, high school. Present results demonstrated how technology skills have evolved in the 21st century among digital natives [14]. Students were confident in their ability to perform all the task included in the TPSA\C-21. This scale is widely used as a measurement tool for research about technology proficiency in 21st century learning. This was important to verify, as it has been found among healthcare students that not all are frequent users of ICT [14]. Furthermore, studies have also purported that Internet use by students was mostly for non-professional related purposes [15][16]. Students in this study felt less confident in some administrative uses of ICT technology and in its use as a tool for their future role as health professionals. Thus, although students seem to have adequate proficiency and confidence, the study also identified some areas in which support, and further development may be required. It points to important issues to be considered in the design and delivery of technology-enhanced curricula in the future.

Although this study provides valuable insights into the technology proficiency and confidence in the ability to perform among dentistry and health professions students, it is not without limitations. The most obvious one was the cross-sectional nature of the study, which precludes a strong conclusion about technology proficiencies amongst the future healthcare workforces. Additionally, this assessment was undertaken using a self-selected sample. This introduces the influences of variation between participants and non-participants in terms of technical self-competency, experience with technology, and other factors. Also, the study relied on self-reported data, which may not be an accurate reflection of the relationship between self-perceived and actual technical competency. However, considering these limitations, we believe that the current approach was adequate given the exploratory nature of the study.

Finally, it points to the different areas of competence that healthcare students must acquire for the use of ICT and Digital Health as a competence [13,17]. These and other developments of ICT, such as with artificial intelligence, robotics, self-learning machines or the need to analyse large amounts of data, will require the development of new competencies among healthcare professionals [13,18].

## V. Conclusions

A telehealth course was designed to provide health professions students with a set of foundational knowledge and skills needed to succeed as a virtual health care practitioner. This study examined the level of technology proficiency (reported as self-assessed 'confidence') amongst the future healthcare workforce. Students in healthcare professional entry-to-practice degrees at the University of Melbourne are technologically proficient, able to use a wide range of technologies regularly in their daily lives. They seem to have adequate proficiency and confidence in their ability to perform all the tasks explored in this evaluation. Nonetheless, the study identified some areas in which support, and further development may be required.

Large-scale testing and validation of the course will follow. However, this initial evaluation provides valuable information, which could be used during the redesign of health science curriculum. It would enable curriculum which meets the needs of students, the healthcare profession, and the community.

## Acknowledgment

This project was funded by a Learning and Teaching Initiative grant and from The University of Melbourne and a Universitas 21 Health Sciences Group International Projects Fund grant. We would also like to acknowledge the students who participated in this study and Dr Diego López, Mr. Luke Davies for their work in this project as assistants, and Prof Alastair Sloan and Prof Linda Denehy for their support, advice, and feedback through this project. Finally, our project advisory committee.

## References

[1] E. R. Dorsey and E. J. Topol, "State of Telehealth," N Engl J Med, vol. 375, pp. 1400, 2016.

[2] M. W. L. Moreira, J. J., P. C. Rodrigues, V. Korotaev, J. Al-Muhtadi and N. Kumar, "A Comprehensive Review on Smart Decision Support Systems for Health Care," in *IEEE Systems Journal*, vol. 13, no. 3, pp. 3536-3545, Sept. 2019, doi: 10.1109/JSYST.2018.2890121T.

[3] Greenhalgh, J Wherton, S. Shaw, and C. Morrison, "Video Consultations for Covid-19," BMJ; vol. 368, m998, 2020.

[4] D. Garcia-Huidobro, S. Rivera, S. V. Chang, P. Bravo, and D. Capurro, "System-wide accelerated implementation of telemedicine in response to COVID-19: mixed methods evaluation," J Med Internet Res, vol. 22:e22146, 2020.

[5] M. Alcocer, et al. "Teaching Telemedicine: The Next Frontier for Medical Educators," JMIR Med Educ, vol. 7, e29099, 2021

[6] S. Edirippulige, and N. R. Armfield, "Education and Training to Support the Use of Clinical Telehealth: A Review of The Literature," J Telemed Telecare, vol. 23, pp. 273-282, 2017.

[7] M. Merolli, D. Capurro, L Davies, D. Lopez, and R. Mariño, "Telehealth Education for Entry to Practice Health Professional Students". Available from: https://mdhs.unimelb.edu.au/engage/learning-and-teaching/2021-learning-and-teaching-symposium#thursday-program/ [retrived: March 2022].

[8] R. Christensen and G. Knezek, "Validating the Technology Proficiency Self-Assessment Questionnaire for 21st Century Learning (TPSA C-21)," J. Digit Learn Teach Educ, vol. 33, pp. 20-31, 2017.

[9] J. Tavares and T. Oliveira, "Electronic Health Record Patient Portal Adoption by Health Care Consumers: An Acceptance Model and Survey," J Med Internet Res, vol.18:e49, 2016.

[10] V. Venkatesh, J.Y.L. Thong, and X. Xu, "Consumer Acceptance and Use of Information Technology: Extending the Unified Theory of Acceptance and Use of Technology," MIS Quarterly vol. 36, pp. 157–78, 2012.

[11] K. Tamilmani, N.P. Rana, S. Fosso Wamba, and R. Dwivedi, "The extended Unified Theory of Acceptance and Use of Technology (UTAUT2): A systematic literature review and theory evaluation," Int J Inf Manage, vol. 57, 2021, doi.org/10.1016/j.ijinfomgt.2020.102269

[12] J. Cohen, Statistical Power Analysis for the Behavioral Sciences. 2nd ed., Hillsdale, NJ: Lawrence Erlbaum publishers, 1988.

[13] M. Brunner, et al. "An eHealth Capabilities Framework for Graduates and Health Professionals: Mixed-Methods Study," J Med Internet Res, vol. 20, e10229, 2018.

[14] A.E. Cowey and H. Potts, "What Can We Learn From Second Generation Digital Natives? A Qualitative Study of Undergraduates' Views of Digital Health at One London University," Proceedings of the Human Factors and Ergonomics Society Annual Meeting, pp2102-2106, 2018.

[15] R. Mariño, E. Habibi, W. Au-Yeung and M. Morgan, "Use of Communication and Information Technology among Victorian and South Australian Oral Health Profession Students," J Dent Educ,vol, 76, pp. 1667-74, 2012.

[16] G. Kennedy, et al., Questioning the Net Generation: A Collaborative Project in Australian Higher Education. In: Who's Learning? Whose Technology? Proceedings ASCILITE 2006. Sydney: Sydney University Press, pp. 413-417, 2006.

[17] R. Mariño, C. Delany, K. Reid, et al., Investigation of the Preparation for Practice of Newly Qualified Dental Practitioners in Australia. Final Report. Melbourne Dental School, The University of Melbourne. 2021. Available from: https://www.adc.org.au/files/research/Preparation_for_Practice_of_Newly_Qualified_Dental_Practitioners_in_Austraila.pdf [retrived: March 2022].

[18] O. Fejerskov, S. Uribe and R., Mariño, Dentistry in a Historical Perspective and a Likely Future of the Profession. In: Mariño R, Morgan M, Walmsley D. (Eds.) Career Paths in Oral Health. Springer International Publishing. pp. 3-19. 2018.

# Medicinal Product Data Standardisation – Prerequisite for Efficient Data Exchange Between Stakeholders and Impact on the (Inter)National Health Systems

## Medicinal Product Data Standardisation in the Agency for Medicinal Products and Medical Devices (HALMED)

Sanja Grčić Plečko[1, 2]
[1]Agency for Medicinal Products and Medical Devices (HALMED), Zagreb, Croatia
[2]Faculty of Information Studies in Novo Mesto, Novo Mesto, Slovenia
e-mail: Sanja.GrcicPlecko@halmed.hr

Maja Fatiga[1, 3]
[1]Agency for Medicinal Products and Medical Devices (HALMED), Zagreb, Croatia
[3]University of Zagreb, Faculty of Organisation and Informatics, Varaždin, Croatia
e-mail: Maja.Fatiga@halmed.hr

Dubravka Sudić[1]
[1]Agency for Medicinal Products and Medical Devices (HALMED), Zagreb, Croatia
e-mail: Dubravka.Sudic@halmed.hr

*Abstract* - **The authors describe challenges and learnings in the process of refactoring internal system for tracking the records of marketing authorization procedures and medicinal product database in the Agency for Medicinal Products and Medical Devices (HALMED) in order to comply with ISO IDMP (International Organization for Standardization, Identification of Medicinal Products) standards. HALMED's plans include the medicinal products data model reconstruction, user interface adaptation, changes on the synchronization processes and establishing connection with SPOR (Substances, Products, Organisation and Referentials) data management services. In a response to the introduction of the ISO IDMP set of standards, the EMA (European Medicines Agency) initiated the SPOR data management service project, with the objective to provide the framework for standardization and structured data in medicinal products description.**

***Keywords – health informatics; ISO IDMP; SPOR; Substances Management Services; Products Management Services; Organisations Management Services; Referentials Management Services; UNICOM; EMA.***

## I. Introduction

Medicinal products identification is a global challenge: healthcare stakeholders at each medicinal product life-cycle stage are capturing in their information systems different sets of data, using different codebooks, different languages, and even different abbreviations to describe medicinal products. Simple example might be pharmaceutical form described as "Film-coated tablet" that can be presented as "Coated tablet," "Tablet," or even "Tabl." or "Tbl.". The same product might be available in different countries with different dosage strengths and package sizes, under different brand names. Even the same name might identify different products [1]. Suppose a patient has to take a drug prescribed by a doctor in his/her home country while visiting another county, where drug with the same brand name is not available. The question is whether the adequate replacement drug could be dispensed. Re-identification of medicinal products might be challenging due to insufficient data on the prescription or to unreadable data due to the language barrier or just different terminology and types of data used to describe medicinal products. When an inappropriate drug replacement is dispensed, patient's safety might be jeopardized and the worst possible outcome might be adverse drug reactions [2].

Covid-19 pandemic has shown the importance of introducing "univocal global identification (named Pharmaceutical Product IDentifier or PhPID) as foreseen in the ISO/CEN suite of IDMP standards. The requirement is supported for all medicinal products by the FDA (U.S. Food and Drug Administration) and by EMA, and facilitated by the EU Innovation Project UNICOM" [3].

EC Regulation (EU) 520/2012 (articles 25 & 26) [4] is addressing the above-described challenges by requiring national competent authorities in EU and Marketing Authorization Holders (MAHs) to apply the ISO IDMP standards for identification of medicinal products.

ISO IDMP standards were introduced with the goal to have a standardized set of drug information across the world, across regulatory and medicinal communities, with the aim to fulfil a need in much wider health care areas. The standards were developed by the International Organization for Standardization (ISO) in cooperation with the International Council for the Harmonization of Technical Requirements for Pharmaceutical Products for Human (ICH), Health level Seven (HL7) and many other international stakeholders and experts [5][6]. ISO IDMP consists of five *Health Informatics – Identification of Medicinal Products* standards that were initially published in 2012 [7][8][9][10]:

- ISO 11615 — Data elements and structures for the unique identification and exchange of regulated medicinal product information;
- ISO 11616 — Data elements and structures for the unique identification and exchange of regulated pharmaceutical product information;
- ISO 11238 — Data elements and structures for the unique identification and exchange of regulated information on substances;
- ISO 11239 — Data elements and structures for the unique identification and exchange of regulated information on pharmaceutical dose forms, units of presentation, routes of administration and packaging;
- ISO 11240 — Data elements and structures for the unique identification and exchange of units of measurement.

Besides preparing an EU IDMP Implementation guide [13], EMA has established the SPOR data management services (***S**ubstances, **P**roducts, **O**rganisation* and ***R**eferentials)* [11][12]. These management services should ensure that all EU stakeholders are using the same value sets for substances, organisations and referential data (*dose forms, units of measurement, units of presentation, routes of administration)* [14].

Although EMA provided an EU Implementation guide [15], complying with the EC Regulation (EU) 520/2012 (articles 25 & 26) requires the national agencies to design a strategic plan for data management. Such a plan requires significant resources in terms of time, subject matter experts and financial resources. [16] Besides performing a detailed analysis of internal business processes and the assessment of data required by other healthcare stakeholders in order to get the required set of medicinal product data, other important tasks need to be fulfilled: the current data model used in medicinal product database should be evaluated, user interface reconstruction should be considered and data must be corrected, standardized and mapped or replaced with the terms and codes of the SPOR value sets.

In this article we describe the approach and experience of the HALMED in refactoring its information system in order to comply with ISO IDMP standards. We believe it will be instrumental to support other agencies planning their road to ISO IDMP compliance and will also contribute to solve the global challenge of medicinal products identification.

In Section II, we describe the background and challenges related to deficiencies in data model of IT system for marketing authorisation processes and medicinal product database in HALMED. In Section III, we describe the initial step of our project of standardization of medicinal product data in HALMED (defining master data set for medicinal product identification). In Sections IV and V we focus on the analysis of the results and on the plan for the refactoring of the current IT system. Next planned steps and preliminary conclusion are provided in Sections VI and VII.

## II. BACKGROUND AND CHALLENGES

HALMED is the Croatian national competent authority that provides regulatory services pertaining to medicinal products, medical devices, homeopathic medicinal products and certain aspects of veterinary medicinal products in accordance with the legislation of the Republic of Croatia.

In 2010, HALMED started to build its own IT system to support marketing authorization processes and its medicinal products database, called NRL (abbreviation for Nacionalni Registar Lijekova, i.e., National Medicinal Products Registry). It was decided that the medicinal products database should be based on the data model described in RDM 3.0 model (Reference Data Model published by EMA), that was confirmed as good practice at that time. As the original RDM data model has a lower sophistication level then the ISO IDMP data model, medicinal products are not described with the level of granularity expected for data exchange in the future EU medicinal products database. For instance, the current data are insufficient for cross-border e-prescription exchange. Currently the medicinal product packaging is described with just a couple of descriptive text fields, and not with the structured data described in ISO 11615 standard [17], on the base of which the EU Implementation guide [18] has been prepared.

Consequences: data exchange with other healthcare stakeholders in the country or abroad is either not possible or very limited or needs to be performed manually. Even internal processes that rely on medicinal products data are missing essential information. For example, data that describe medicinal product packaging is essential for the process of calculating maximal wholesale price based on the comparable medicinal products' prices from reference countries. Each medicinal product packaging from a reference country could be considered the same as domestic packaging if it has the same active substance, the same strength, the same pharmaceutical dose form [33][34], and the same pack size as the domestic packaging. From the listed attributes, the biggest challenge is to compare the packages themselves, especially while in the medicinal products database (that is used for list of domestic products) in internal system called NRL-PKL-PhV (PKL stands for Provjera Kakvoće Lijekova, i.e., Medicinal Products Quality Control and PhV stands for Pharmacovigilance), packages are described with just two descriptive text fields. There is no data describing the container, its volume and whether the vial is for single or multiple use. Standardized description of units of presentation is missing, as is the quantity of pharmaceutical product in the container, the number of containers in package, etc., etc.

### A. *Description of the IT system at HALMED for core business processes and medicinal product database*

NRL system has been developed as a Web-based application that consists of a database, a Web interface, and Web services for integration with other internal IT systems. Although HALMED's IT systems have been developed and upgraded in phases over the years, they are all interconnected with the services for data exchange and automation of process tasks, so that users are enabled to carry out all steps required for a business process via only one application – their core business process application. All process documents are stored on DAIS (Digitalni Arhivski Informatički Sustav, i.e., Digital Archival Information System) [36][37], HALMED's Enterprise Content Management System built on IBM FileNet, and accessed directly from NRL. HALMED used to receive marketing authorisation applications documentation in paper form and those documents were then digitized, enriched with metadata, stored on DAIS and made accessible directly through the NRL application. Through Web services, the system is also connected with the other HALMED's systems such as the filing system (Centrix), Archival Management System (Pismohrana) and invoicing system, so users can complete all their tasks within one interface.

Over the years, the NRL system, built to support medicinal products marketing authorization processes, has grown and even became integrated with two other internal systems that rely on the same medicinal products data into one system called NRL-PKL-PhV.

That system supports core business processes:

- Case tracking for marketing authorization procedures (tracking of all procedural phases, tracking of deadlines and tasks completed by assessors, supporting the Committee for Medicinal Products processes, business reporting).
- Pharmacovigilance tasks (Risk Minimization Measures (RMM) and Referral procedures).
- Inspectorate activities related to planning and executing medicinal products sampling, quality control of medicinal products (Human and Veterinary) in the Official Medicines Control Laboratory (OMCL) (filing incoming samples, sample analysis and analysis task assignment, reagent management and management of standards, reporting of results and filing outgoing documents) [19].

NRL-PKL-PhV system (as shown in Figure 1) consists of three applications that share the same user interface, database, document store on DAIS, administration tools, and codebooks (referential lists). The latter are daily synchronized with SPOR RMS (Referentials Management Services) [20][21] by using the Application Programming Interface (API) lists and EUTCT (European Union Telematics Controlled Terms) substance list.

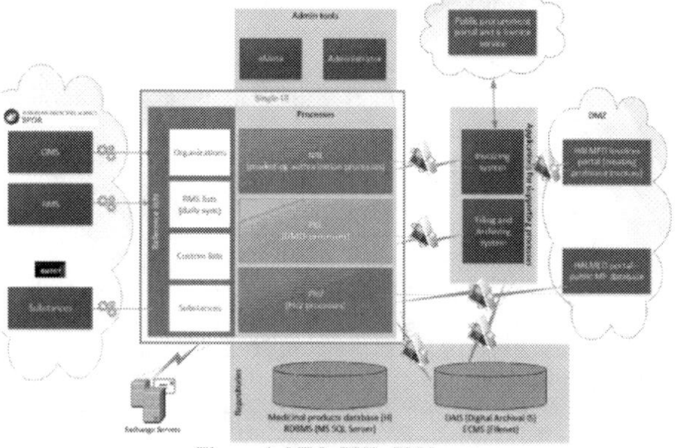

Figure 1. NRL-PKL-PhV system

In order to be able to capture all necessary data describing medicinal products, HALMED had to decide on the best pathway to ensure full compliance with ISO IDMP standards: either by building a new system up from the scratch or either by refactoring the old NRL system. HALMED considered on one hand that medicinal product data stored in NRL-PKL-PhV were not yet exchanged with other national-wide eHealth systems and on the other hand that a number of very complex processes were already supported by the system developed over the years. HALMED decided thus to opt for a refactoring of the NRL-PKL-PhV system, resulting in the reconstruction of the data model, the adaptation of the user interface, and the modification of the synchronisation processes.

With the decision to undertake a NRL-PKL-PhV system refactoring, new questions arose about the complexity and the quality of the data used up to then.

## III. Analysis of the Master data set and reconstruction of the data model

In order to better understand the business needs, all the processes that rely on the medicinal product data were thoroughly analysed. Besides the NRL-PKL-PhV system, that is supporting above mentioned processes, the following additional processes utilize medicinal products data:

- Calculating the maximal wholesale price of medicinal products,
- Granting authorizations for parallel imports of medicinal products,
- Giving approval for entry and importation of medicinal products and for emergency entry and importation of medicinal products.

The Master data set for medicinal product was defined after a number of workshops with subject matter experts, resulting in a thorough analysis of internal processes that generate or utilize medicinal products data. In addition, requests from external consumers of medicinal products data were analysed.

When defining the master data set, we also analysed the data models described in ISO IDMP standards. First and the most frequently used in our analysis was the ISO 11615:2017 for the unique identification of regulated medicinal product information [17]. When it came to questions on how to describe complex compositions of pharmaceutical products, ISO 11238:2018 [22] was studied for deeper understanding of Substances identification data and ISO 11616:2017 [23] for Pharmaceutical Product information.

During the process of drafting NRL-PKL-PhV system refactoring plan, we obtained very valuable guidance through the EU Implementation Guide v.2.1 [15] where all the medicinal product data that are required by EMA for future PMS (Products Management Services) and medicinal product data exchange are described with full details.

Relying on the results of the analysis of processes utilizing medicinal products data, a detailed assessment of current data model was performed, resulting in a thorough comparison with ISO IDMP standards (especially ISO 11615 for medicinal product [17]) in order to detect the gaps. First findings have shown two important refactoring areas:

- Packaged Medicinal Product: Manufactured Item (and its composition) should be introduced into the data model, as well as Packaging Item (Container) and Device
- Pharmaceutical Product needs some adjustments: mandatory specification of the modifiers for chemical substances; introduction of Specified substance (for proteins and biologicals) and Strength (Reference Strength as Presentation and Concentration strength were already implemented).

Figure 2. The methodology of refactoring internal IT system and medicinal product database in order to comply with ISO IDMP standards

After thorough analysis of the current data model, ER (Entity Relationship) diagrams and restructured data models were prepared for Packaged Medicinal Product and for Pharmaceutical Product. In the next section the reconstruction plan for Packaged Medicinal Product is described in more details.

Analysis has shown that the data set defined in the EU Implementation Guide [15] was sufficient with the exception of some additions like Specified Substances and some physical characteristics of containers and devices.

The whole process and methodology of refactoring internal system with medicinal product database, in order to comply with ISO IDMP standards, is shown on Figure 2.

### A. *Reconstruction of data model related to Packaged Medicinal Product*

On Packaged Medicinal Product ER diagram (Figure 3) we colour-coded the entities:

- Red rectangles are those entities that should be introduced in currently implemented data model (Device, Packaging item (container) and Manufactured Item)
- Yellow rectangles would need some adjustments (Packaged Medicinal Product, Pharmaceutical Product, Ingredient),
- Green rectangles are already compliant with ISO standards (Medicinal Product and Substance):

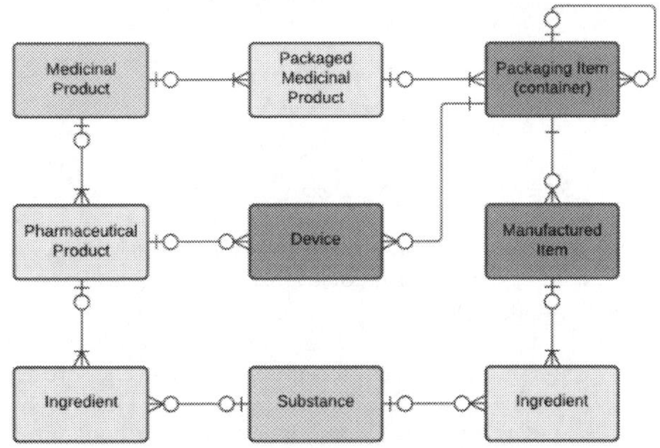


Figure 3. Entity Relationship Diagram: Packaged Medicinal Product

As for the user interface (UI) reconstruction, on tab "Packaging" (both in the module where medicinal product data can be reviewed and in module for case management) changes should be made that enable:

- Input of more Packaged Medicinal Products (i.e., outer packages)
- Input of more Package Items (Containers) with the possibility of copying data from existing Package Items
- Input of Devices with possibility of copying data from existing Devices
- Input of Manufactured Items with possibility of copying data from existing Manufactured Items
- Input of Ingredients for Manufactured Items: Substance and Specified Substance, with the Strengths and Reference Strengths (concentration and presentation)
- Copying/relating Ingredients for Manufactured Items and Ingredients for Pharmaceutical Products.

In Sections III.A.1) - III.A.3) the introduction of new entities is described in more details.

*1) Introduction of a new entity in the data model: Container*

When introducing Container as new entity in the data model, the following aspects need to be considered (Figure 4):

- Container can have zero or more Containers that are referencing it as parent.
- Container can be referenced by different Devices or Manufactured items.
- Attributes that are describing Container: Container Type, Material and Quantity should be added to the data model.

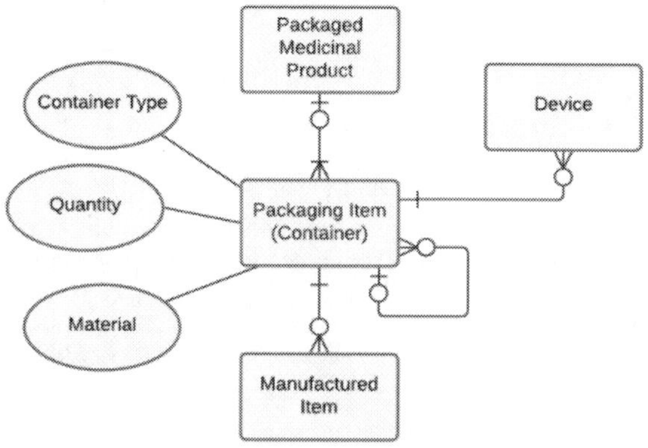


Figure 4. Entity Relationship Diagram: Packaging Item (Container)

*2) Introduction of a new entity in the data model: Device*

There are two possible relationships between Device and Package Item (Figure 5):

- The medical device is integrated and contains the medicinal product for administration (e.g., pre-filled syringes, pre-filled pen). In this case Device is also the primary packaging/package item container for the medicinal product.
- Device as independent element contained within secondary packaging (for example in outer box).

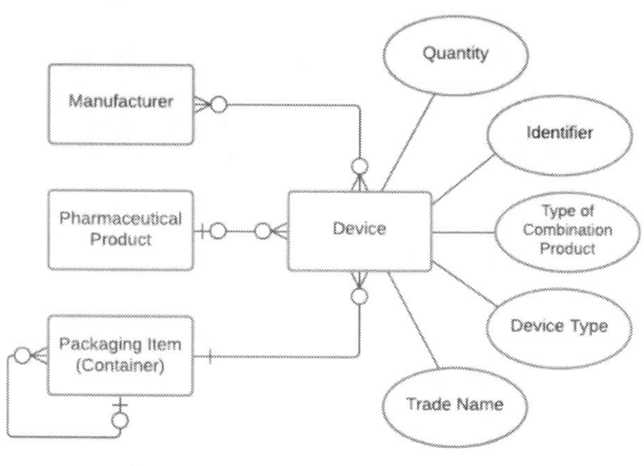


Figure 5. Entity Relationship Diagram: Device

*3) Introduction of a new entity in the data model: Manufactured Item*

Manufactured Item is the product as it is authorised and before transformation into the administrable pharmaceutical form [18].

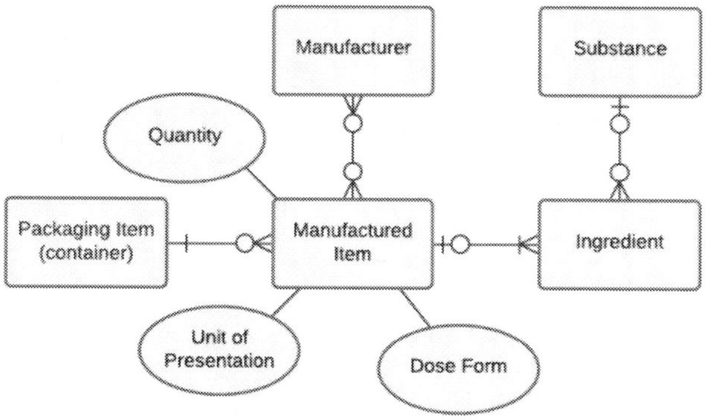


Figure 6. Entity Relationship Diagram: Manufactured item

When introducing Manufactured Item as new entity in the data model (Figure 6), following attributes that are describing Manufactured Item should be introduced: Dose Form, Unit of presentation, Quantity.

### *B. Reconstruction of the data model related to Pharmaceutical Product*

In the Pharmaceutical Product entity relationship diagram (Figure 7), red rectangles are those entities that need to be introduced in the currently implemented data model (Manufactured Item, Device, Strength (Presentation, Concentration)); yellow ones are those that need some transformation (Pharmaceutical Product, Ingredient, Reference Strength (Presentation, Concentration)) and green ones are those that don't need any changes (Medicinal Product, Substance and Administrable Dose form).

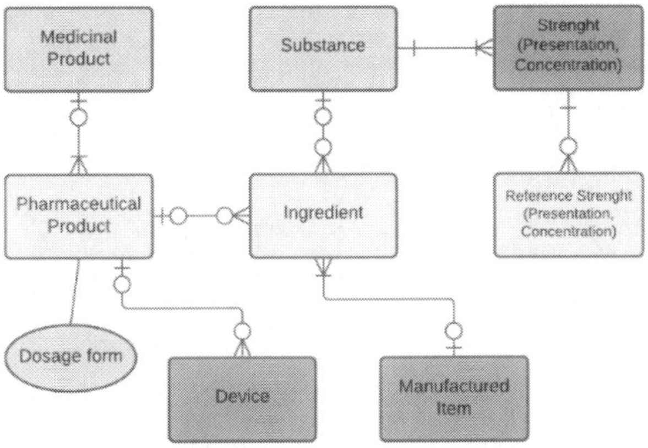


Figure 7. Entity Relationship Diagram: Pharmaceutical Product

Reconstruction activities related to Pharmaceutical Product should include the introduction of new entities in the data model: Specified Substance; Strength (presentation) and Strength (concentration); also, the user interface needed reconstruction so that Strength (presentation) and Strength (concentration) could be recorded, as well as the Specified Substances. Reference strength (presentation) and Reference strength (concentration) needed to be migrated to new database tables.

The structure of related tables currently enables data entry for the majority of entities, but does not (fully) comply with ISO IDMP data model. The Ingredients table should be thoroughly analysed and, based on findings, new table structure and relations should be defined, as well as data migration.

Pharmaceutical Product name/description is necessary when medicinal product has more than one Pharmaceutical Product and currently, in many cases, this is not recorded in the correct form (physical characteristics plus units of presentation).

In the next section of this paper, the transition to SPOR RMS referential lists for internal codebooks is described.

## IV. Introduction of RMS Referential Lists

In order to ensure data consistency in describing medicinal products, common referential lists (codebooks) should be used. Fortunately, HALMED has been using EUTCT lists [24] for internal codebooks from the early stages of NRL system development and those codebooks were synchronized on a daily basis. Table 1 shows the EUTCT lists that were used in marketing authorisation processes and related NRL lists.

Until recently, it was allowed to add user-defined terms in NRL-PKL-PhV codebooks. Planning for transition to RMS lists requires assessment of different scenarios:

- Codebooks with EUTCT terms and no custom-added terms
- Codebooks based on EUTCT lists and with user added terms
- Custom codebooks in NRL-PKL-PhV system
- Introduction of new RMS codebooks to NRL-PKL-PhV system (no codebooks were used)

Table 1 EUTCT lists that were used in marketing authorisation processes

| EUTCT list | NRL codebook |
|---|---|
| Substance (still in usage) | Tvari |
| Country | Države |
| Dosage Form Category | Grupe farmaceutskih oblika |
| Dosage Form | Farmaceutski Oblici |
| Units of Measurement | Jedinice mjere |
| Route of Administration | Put primjene |
| Authorisation Status | Registracijski status lijeka |
| Marketing Status | Status lijeka na tržištu |
| Ingredient Role | Vrste sastojaka |
| Container | Vrste pakiranja |
| Supply | Mjesto izdavanja |
| Legal Status for the Supply | Način propisivanja |
| Special Precaution for Storage | Uvjeti čuvanja |
| Quantity Operator | Operatori količine |

Different scenarios related to the usage of the codebooks and the associated implementation steps are described in Sections IV.A - IV.D. In all the described scenarios, we consider an automatic daily synchronisation with the Referentials Management Services via the Application Programming Interface (API).

### *A. Codebooks with EUTCT terms and no custom-added terms*

In the case of internal codebooks that were synchronized with EUTCT lists (no longer allowing custom-added additional terms), transition to RMS lists started with the detection of corresponding items in the RMS SPOR list. In the NRL-PKL-PhV system, changes to the database (adding new fields for RMS list) and user interface were made. In order to keep the record of the change of localized terms, term versioning was introduced, and likewise for the three other scenarios, as described in Section IV.E.

### *B. Codebooks based on EUTCT lists and with user-added terms*

For user-added terms, we performed a data cleansing, deactivating user-added terms and disabling the possibility of adding new custom terms in the codebook. After detecting the corresponding RMS list, all necessary changes were made in database and user interface. For data cleansing and for mapping old terms to RMS terms, a custom-made administration tool was used.

### *C. Custom codebooks*

Before connecting to corresponding RMS list, data quality analysis and data cleansing were performed; all the changes in database structure and user interface were also completed so that the new RMS list could be imported and custom (old) terms mapped with RMS terms and redundant terms could be deactivated.

*D. Introduction of new RMS codebooks to NRL-PKL-PhV system*

Business experts with the IT team are continuously assessing the RMS referential lists and are prioritizing their adoption in NRL-PKL-PhV system. Reconstruction of the database and changes in system functionalities were only introduced once a full agreement had been reached on the final RMS list.

During the data model reconstruction, introduction of new RMS lists should be planned, based on the entity's attributes.

*E. Term versioning in internal codebooks*

With the introduction of RMS lists, SPOR and NRL term versions are stored separately in internal codebook. That intermediary step is needed as the old terms cannot be changed before the product information documents are officially changed. This requires a new submission by the Marketing Authorisation Holder. Until then, old terms remain intact and NRL users are being informed about the term change via an e-mail notification (Figure 8). Moreover, users are also notified in the user interface: on the status bar of the opened case and/or medicinal product (Figure 9) and on the list of terms in the field where it appears (Figure 10). Besides, in the search engine both old and new term versions are detectable.

NRL Nacionalni registar lijekova
NRL Nacionalni registar lijekova;
Production -- SPOR sinkronizacija - razlike

**Farmaceutski oblici**

| Promijenjen je zapis: 200000025182 | | |
|---|---|---|
| Naziv kolone | Stara vrijednost | Nova vrijednost |
| Version | 2 | 4 |
| ValidFrom | 2022-02-16 14:14:20.626 | 2022-05-24 15:31:43.959 |
| SporStatusID | NULLIFIED | NON_CURRENT |
| Name_EN | Emulsion and solution for emulsion for injection | Solution and emulsion for emulsion for injection |

Figure 8. Localized term versioning – e-mail notifications

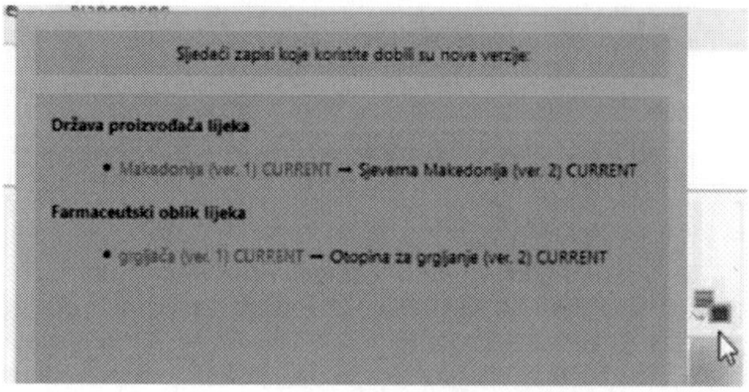


Figure 9. New term indicated on status bar

Old terms are displayed in red letters when they appear in the user interface and in drop down lists. In marketing authorization application for a new medicinal product, only new terms will be shown on all lists (Figure 10).

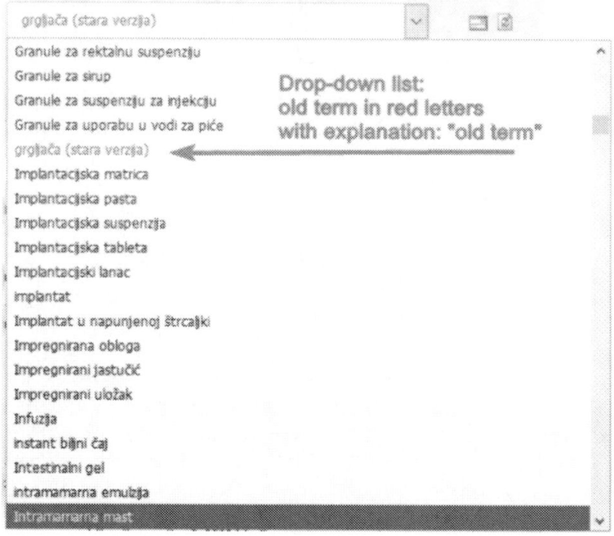


Figure 10. Indication of old term version in use

## V. USER INTERFACE RECONSTRUCTION

In following paragraphs, proposed changes in user interface are described.

*A. Packaged Medicinal Product*

During the marketing authorisation procedure HALMED is allocating a unique Marketing Authorisation Number to a Medicinal Product and that number will not change during the whole Medicinal Product life-cycle, regardless of changes on the product name, marketing authorisation holder or any other data. Furthermore, all packaging's of the same medicinal product are designated with a packaging digit code (from 01 to 99) and with the Marketing Authorisation Number of the Medicinal Product, followed by the packaging digit code. These elements together are forming the Marketing Authorisation Number for Packaging. In the NRL-PKL-PhV system, all packaging of the same medicinal product are listed in one tab in the module *Medicinal product list (Središnji podaci o lijekovima)* and in the module *Processing applications (Obrada predmeta)*. Currently in the "Packaging" only a few descriptive text fields are available so this tab should be reconstructed in order to enable entering all data for Packaged Medicinal product.

The proposed solution is to introduce a hierarchical tree-list grid (Figure 11) for the presentation of all the levels of Medicinal Product Package data:

- from outer package
- containers inside outer package,
- containers inside containers,
- devices that might be included in outer package or in container,
- integral or co-packed device.

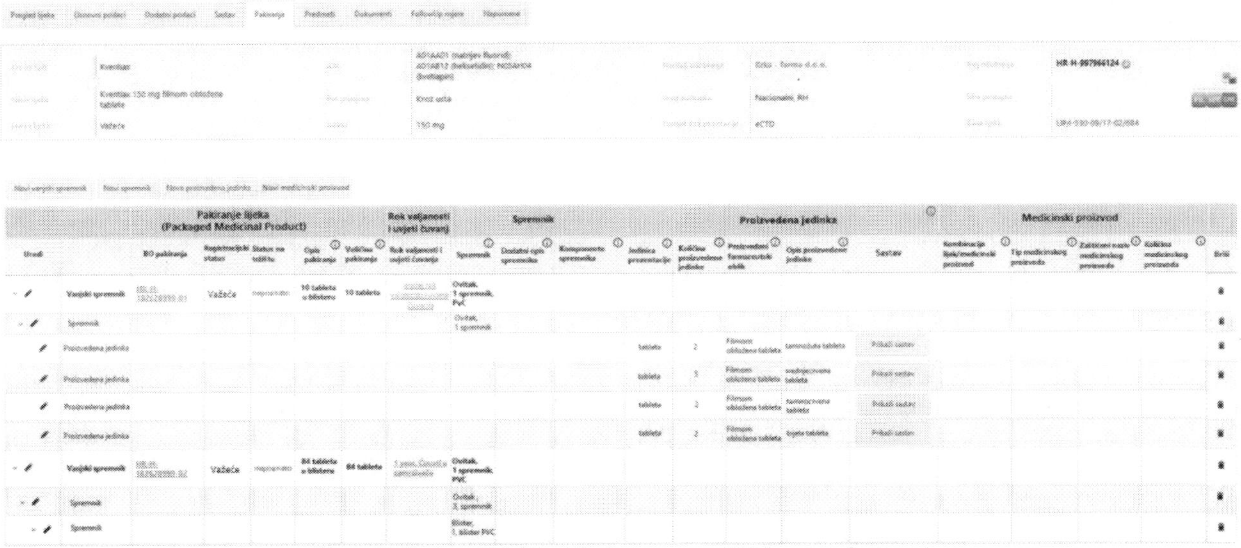

Figure 11. Packaged Medicinal Product – proposed UI (user interface) mock-up

The tree-list ends on the level of immediate container with details about the manufactured item and its composition. Furthermore, all additional attributes like shelf life and special precautions for storage will be visible in the grid, often with concatenated values from more fields that were used to describe the entity. For example, for the presentation strength, all data entered in separate fields (fields: Quantity operator, Strength (Presentation single value or low limit) - numerator and denominator, Strength (Presentation high limit) - numerator and denominator) in the pop-up will be presented in one field on grid Manufactured Item composition Strength (Presentation) as shown on Figure 12.

Adding data for new outer packages, containers, devices and manufactured items will be enabled with new buttons positioned above the grid. For editing, the pen symbol on the left edge of the grid opens the same pop-ups. Each pop-up consists of text fields, such as fields for the description of outer packages or manufactured items. In addition, there are fields with controlled number values and fields controlled by drop down lists where only the attributes from RMS reference lists would be available for selection.

While the Medicinal Product could have more packaging's that might be similar in their components, the user will be able to copy data from one outer packaging / container / device / manufactured item to the other in order to save time for entering all the required details.

Composition of Pharmaceutical Product is currently presented via the tab "Ingredients," where Substances and Reference Strengths (concentration and presentation) are recorded. In the reconstructed user interface, adding and editing Specified Substances and Strengths (concentration and presentation) are enabled and a very similar solution will be implemented for capturing Manufactured Item composition.

Dialog boxes for adding and editing information about the Manufactured item and Pharmaceutical product composition facilitate the definition of the Ingredient role (for e.g., substance (or ingredient) with the role of precise active substance, excipient…), adding Substance and/or Specified Substance and describing their Strength (Presentation and Concentration) and Reference Strength (Presentation and Concentration) (Figure 13).

Figure 12. UI mock-up: Manufactured item composition

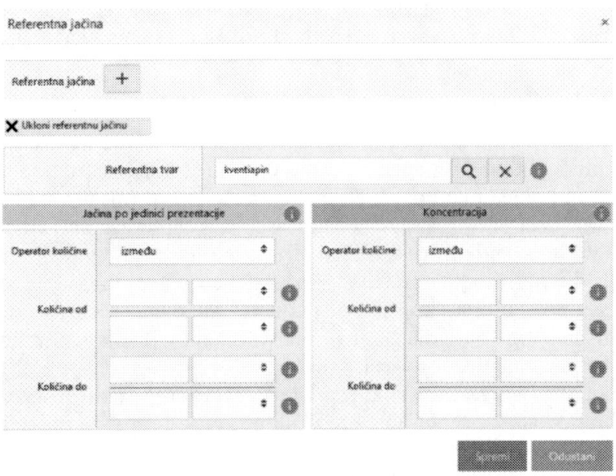


Figure 13. UI mock-up for Manufactured item composition editing pop-ups (adding reference substance and strength)

For a significant number of medicinal products, the Pharmaceutical Product composition and the Manufactured item composition would be the same and the plan is to enable copying composition data for some or all substances and strengths from one to another.

## VI. FURTHER WORK

The work presented in this document is just the beginning of the refactoring of the current system, step one on the Figure 14. This refactoring is relying heavily on the assessment of internal business processes which are generating or utilizing medicinal product data and on the thorough analysis of the gaps in the data model of the current system, in comparison with the data model provided by the ISO IDMP standards. Besides data model reconstruction and user interface adaptation, connection to the new RMS referential lists that are used to describe medicinal product packages should be established; and this is also needed for SMS (Substances Management Services) and future PMS (Products Management Services). In order to make this effective, applications that will send and receive data from future PMS and which receive substance data from SMS, will be programmed using FHIR (Fast Healthcare Interoperability Resources). [25][26][27][28] Moreover, web services will be built to send data from the HALMED database to the national medicinal product dictionary database that is part of the EU funded project eLijekovi [32].

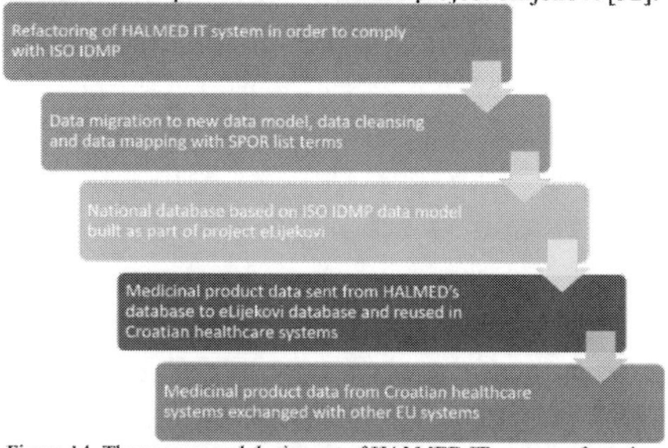

Figure 14. The process and the impact of HALMED IT system refactoring on national and EU wide healthcare systems

## VII. CONCLUSION

The complexity of the processes related to the marketing authorisation of medicinal products and to the production of reliable medicinal product data is immense. It implies taking into account different healthcare stakeholders and their activities and processes that are utilizing data in different medicinal product life-cycle stages. The more obvious ones are prescribing, capturing patients' medical records and dispensing but there are many others. It also implies dispensing the products that are prescribed in another country where the same product is registered with different brand name, or the same brand name is used for a completely different medicinal product.

The possibility of uniquely identifying medicinal products globally [29] is becoming top priority, as was demonstrated by the need of tracking the global use of COVID-19 vaccines during the pandemic [3]. The cross-border mobility scenarios where e-prescriptions and patient summaries would be easily exchanged between countries offer a unique opportunity of getting closer to this ultimate goal, with direct positive impact on patients' safety[3][30].

With the introduction of the ISO IDMP set of standards, EMA initiated the SPOR project to define four domains of master data in regulatory processes: Substance, Products, Organisations and Referentials [9]. RMS lists and OMS (Organisations Management Services) organisations registry are ready to use while SMS and PMS are expected to become available soon for substance and medicinal products data exchange. The UNICOM project (funded by the European Union's Horizon 2020 research and innovation program under grant agreement No 875299) is an essential accelerator of the ISO IDMP standards implementation. Its ambition is to improve patient safety by ensuring that "any medicine and what it contains can be accurately identified anywhere in the world." The project gathers 40 partners representing all the actors of the value chain. 11 participating National Medicines Authorities are concretely experimenting ISO IDMP implementation; together they build a common understanding of what needs to be done towards successfully implementing IDMP, and how to learn from each other and share best practices. [30].

Nevertheless, the implementation process is very demanding for the industry, the healthcare organisations or the regulators. It requires a deep assessment of all internal business processes that rely on medicinal product data and an analysis of external stakeholder's dataset needs so that a master data set can be defined; Organisations need to become acquainted with the ISO IDMP data models and the EU IDMP Implementation guide [13][31]. One of the first decisions to be made is whether an entirely new system (off-the-shelf or custom made) needs to be implemented or whether the old system can be refactored. HALMED opted for the refactoring of the existing system NRL-PKL-PhV. It consists of three distinct applications that are sharing the same medicinal products database, same document repository, referential lists, administration tools and integration services to other HALMED's information systems. Significant adaptations of the user interface are also needed in order to enable entering and reviewing medicinal product data related to Packaging, Manufactured item and its composition and the Pharmaceutical Product composition. Once the old data model and the user interface reconstruction will be fully completed and the new RMS referential lists introduced, HALMED will then be able to exchange all data required by the EU IDMP Implementation guide with the Product Management Services. Moreover, as part of the EU funded project eLijekovi, the future national medicinal product database (that will be foundation for all the future eHealth services consuming medicinal products) will be built in accordance with the ISO IDMP set of standards. By complying with ISO IDMP standards, the national medicinal product database will significantly impact the Croatian eHealth system in general. It will provide an essential contribution to data interoperability throughout the Croatian healthcare systems and therefore enable a meaningful and quality data exchange between all stakeholders.


## Acknowledgment

Authors presented challenges and learnings from ISO IDMP standards implementation project at the Agency for Medicinal Products and Medical Devices of Croatia (HALMED). The project is funded by HALMED and is partially supported by the UNICOM project that has received funding from the European Union's Horizon 2020 research and innovation programme under grant agreement No. 875299 [30]; R. H. Vander Stichele (I~HD) and L. Nicolas (EHTEL) have kindly accepted to review this paper.


## References


[1] K. A. Stroetmann, "The univocal identification and safe dispensation of medicinal products across Europe – challenges and solution proposal," MS Med Inform Biom Epidemiol 2017, vol. 13, no. 2, pp. 1–15, 2017.

[2] K. A. Stroetmann, "Meeting the semantic challenge of the globally unique identification of medicinal products - the openMedicine approach," Stud. Health Technol. Inform., vol. 209, no. January, pp. 170–174, 2015.

[3] R. H. Vander Stichele, C. Hay, M. Fladvad, M. C. J. M. Sturkenboom, and R. T. Chen, "How to ensure we can track and trace global use of COVID-19 vaccines?," Vaccine, vol. 39, no. 2, pp. 176–179, 2021.

[4] European Commission, "Implementing Regulation (EU) 520/2012," Off. J. Eur. Union, vol. 2012, no. 520, pp. 5–25, 2012.

[5] K. Koshechkin, G. Lebedev, and F. Eduard, "Implementation of IDMP standards as a means of creating a unified information space in the field of drug circulation," Procedia Comput. Sci., vol. 176, pp. 1745–1753, 2020.

[6] W. Goossen, J. T. Costa, C. Hay, M. Melgara, and R. VanderStichele, "Deliverable 1.3: Initial openMedicine infostructure," Project openMedicine, Available from EC Research & Innovation Participant portal: https://ec.europa.eu/research/participants/documents/downloadPublic?documentIds=080166e5a98384a6&appId=PPGMS.

[7] M. Anelli, M. Dollen, X. Fournie, E. Ghanem, and Z. Koberidze, "Pharmacovigilance in 2020: boldly shaping the future - an overview Part 2: Identification of Medicinal Products (IDMP) implementation," EUCROF Pharmacovigilance Working Group, pp. 1–12, 2018.

[8] European Medicines Agency, "Products Management Services - Implementation of International Organization for Standardization (ISO) standards for the identification of medicinal products (IDMP) in Europe: EU Implementation guide Version 2.1 - Introduction," 2021.

[9] European Medicines Agency, "Data on medicines (ISO IDMP standards): Overview" [Online]. Available: https://www.ema.europa.eu/en/human-regulatory/overview/data-medicines-iso-idmp-standards-overview. [retrieved: March, 2022].

[10] S. Tranchard, "Revised IDMP standards to improve description of medicinal products worldwide" [Online]. Available: https://www.iso.org/news/ref2234.html. [retrieved: March, 2022].

[11] European Medicines Agency, "European Medicines Agency (EMA) master data management roadmap Substance, Product, Organisation and Referential data" EMA/730453/2014. London: European Medicines Agency; 23 April 2015. Available: http://www.ema.europa.eu/docs/en_GB/document_library/Other/2015/04/WC500186290.pdf.

[12] European Medicines Agency, "Substance, Product, Organisation and Referential (SPOR) master data." [Online]. Available: https://www.ema.europa.eu/en/human-regulatory/research-development/data-medicines-iso-idmp-standards/substance-product-organisation-referential-spor-master-data. [retrieved: March, 2022].

[13] "EU IDMP Implementation guide - version 2.1." [Online]. Available: https://www.ema.europa.eu/en/human-regulatory/research-development/data-medicines-iso-idmp-standards/spor-master-data/substance-product-data-management-services#eu-idmp-implementation-guide---version-2.1-section. [retrieved: March, 2022].

[14] R. Munnik, "The clock is ticking on European data standards for medicinal products," Life Sci. Lead., vol. 2, pp. 2–5, 2021.

[15] European Medicines Agency, "EU IDMP Implementation guide - version 2.1." [Online]. Available: https://www.ema.europa.eu/en/human-regulatory/research-development/data-medicines-iso-idmp-standards/spor-master-data/substance-product-data-management- services#eu-idmp-implementation-guide---version-2.1-section. [retrieved: March, 2022].

[16] M. Kloft, A. Herrmann, and J. Werner, "Identification of medicinal products - how to turn a compliance project into a strategic initiative with additional benefits." [Online]. Available: https://globalforum.diaglobal.org/issue/november-2019/identification-of-medicinal-products/. [retrieved: April, 2022].

[17] International Organization for Standardization, "ISO 11615:2017, Health informatics — Identification of medicinal products — Data elements and structures for the unique identification and exchange of regulated Medicinal Product information," 2017.

[18] European Medicines Agency, "Products Management Services - Implementation of ISO standards for the identification of medicinal products (IDMP) in Europe: EU Implementation guide v.2.1 - Chapter 2: Data elements for the electronic submission of information on medicinal products." [Online]. Available: https://www.ema.europa.eu/en/human-regulatory/research-development/data-medicines-iso-idmp-standards/spor-master-data/substance-product-data-management-services. [retrieved: March, 2022].

[19] HALMED, "Agency for Medicinal Products and Medical Devices of Croatia. About HALMED." [Online]. Available: https://www.halmed.hr/en/O-HALMED-u/Osnovni-podaci-i-dokumenti/Djelatnosti/. [retrieved: March, 2022].

[20] European Medicines Agency, "Referentials Management Services (RMS)." [Online]. Available: https://www.ema.europa.eu/en/human-regulatory/research-development/data-medicines-iso-idmp-standards/spor-master-data/referentials-management-service-rms. [retrieved: March, 2022].

[21] European Medicines Agency, "RMS (Referentials Management Services)." [Online]. Available: https://spor.ema.europa.eu/rmswi/#/. [retrieved: March, 2022].

[22] International Organization for Standardization, "ISO 11238:2018, Health informatics - Identification of medicinal products - Data elements and structures for the unique identification and exchange of regulated information on substances," 2018.

[23] International Organization for Standardization, "ISO 11616:2017 - Health informatics - Identification of medicinal products - Data elements and structures for the unique identification and exchange of regulated Pharmaceutical Product information," 2017.

[24] "EUTCT lists." [Online]. Available: http://eutct.ema.europa.eu/eutct/showAvailableListsDisplay.do?guestuser=true. [retrieved: March, 2022].

[25] HL7, "Fast Healthcare Interoperability Resources." [Online]. Available: https://www.hl7.org/fhir/. [retrieved: March, 2022].

[26] European Medicines Agency, "Product Management Service (PMS) - Implementation of International Organization for Standardization (ISO) standards for the Identification of Medicinal Products (IDMP) in Europe - Chapter 1: Registration requirements," 2021. [Online]. Available: https://www.ema.europa.eu/en/human-regulatory/research-development/data-medicines-iso-idmp-standards/spor-master-data/substance-product-data-management-services. [retrieved: March, 2022].

[27] European Medicines Agency, "Products Management Services - Implementation of ISO standards for the Identification of Medicinal Products (IDMP) in Europe: EU Implementation guide Version 2.1 - Chapter 6: Technical specifications on structure and format," 2021. [Online]. Available: https://www.ema.europa.eu/en/human-regulatory/research-development/data-medicines-iso-idmp-standards/spor-master-data/substance-product-data-management-services. [retrieved: March, 2022].

[28] European Medicines Agency, "Substances, Products, Organisations, Referentials (SPOR) - SPOR API v2 Specification," February, 2020. [Online]. Available: https://www.ema.europa.eu/en/human-regulatory/research-development/data-medicines-iso-idmp-standards/spor-master-data/substance-product-data-management-services. [retrieved: March, 2022].

[29] V. A. Perkins, "IDMP: an international standard for lifecycle data management," Appl. Clin. Trials, vol. 26, no. 4/5, 2017.

[30] "UNICOM." [Online]. Available: https://unicom-project.eu/. [retrieved: March, 2022].

[31] A. Williams, "Preparing – at last – for IDMP adoption : how life sciences should ensure readiness," Journal for Clinical Studies, vol. 13, no. 3, pp. 16–17, 2022.

[32] "eLijekovi – Integrirani informatički sustav za upravljanje lijekovima (KK.02.2.1.01.0011)" [Online]. Available: https://zdravlje.gov.hr/elijekovi-integrirani-informaticki-sustav-za-upravljanje-lijekovima-kk-02-2-1-01-0011/5394. [retrieved: April, 2022].

[33] R. Vander Stichele and D. Kalra, "Aggregations of substance in virtual drug models based on ISO/CEN standards for Identification of Medicinal Products (IDMP)." Stud Health Technol Inform. 2022 May 25;294:377-381. doi: 10.3233/SHTI220478. PMID: 35612100.

[34] R.H. Vander Stichele, J. Roumier, and D. van Nimwegen, "How granular can a dose form be described? Considering EDQM standard terms for a global terminology." Appl. Sci. 2022, 12, 4337. https://doi.org/10.3390/

[35] „Preparations for eCTD and implementation of digital archival information system“ [Online]. Available: https://www.safu.hr/en/news/preparations-for-ectd-and-implementation-of-digital-archival-information-system. [retrieved: May, 2022].

[36] A. Rajh, T. Karlović, and R. Gospodnetić, "IPA 2009 TAIB: Preparations for eCTD and implementation of Digital Archival Information System", project, 2014. [Online]. Available: https://www.researchgate.net/project/IPA-2009-TAIB-Preparations-for-eCTD-and-implementation-of-Digital-Archival-Information-System. [retrieved: May, 2022].

[37] A. Rajh and H. Stančić, "Planiranje, izgradnja i uspostava digitalnog arhiva (Planning, development and implementation of a digital archive)". Arhivski vjesnik. 51. 41-62. 2010

# Design and Evaluation of a New Nurse-Led Intervention for the Management of Bariatric Surgery Patients

Cláudia Amaro dos Santos,[1,2] João Gregório[1]
1 - CBIOS – Universidade Lusófona's Research Center for Biosciences & Health Technologies
Lisbon, Portugal
2 - Center for Integrated Responsibility Bariatric Surgery and Metabolic Diseases
Évora, Portugal
email: cmendes@hevora.min-saude.pt

*Abstract* — **Obesity has an increasing incidence and bariatric surgery emerges as a treatment for severe and morbid obesity, as well as for its associated diseases, with proven success rates. In this context, patient follow-up by a case manager, who guides the provision of specialized care focused on adapting the patient to the new reality, can prove to be essential to achieve positive results. This study, guided by *the Design Science Research Methodology (DSRM),* will have as main objective to design an intervention [*Case-managing program]* aimed at patients undergoing bariatric surgery. In the evaluation phase of the intervention, an experimental, controlled, and randomized study will be developed, with intervention group and control group. This project aims to be the first study to investigate the effect of specialized interventions on patients who are candidates for bariatric surgery, with evidence based on mixed, *face-to-face, and e-health* programs, on the management and results of bariatric surgery.**

***Keywords- e-health; case manager; perioperative care; bariatric surgery; nurse case management.***

## I. INTRODUCTION

Obesity is considered a 21st century epidemic, with more than 4 million deaths worldwide [1]. It manifests itself as a chronic disease, but also as a risk factor for numerous other pathologies and consequently worse health outcomes. Its prevalence justifies the priority in its treatment. Bariatric surgery is one of the most effective long-term treatments for obesity [2]. The success of bariatric surgery is usually measured by the percentage of weight lost. However, this success is dependent on several factors, such as the preparation for surgery. This path involves multidisciplinary guidance, namely related to health promotion, well-being and self-care, functional readaptation, and pre-habilitation, consistent with patient satisfaction [3].

It is essential to begin health education to prepare patients for surgery, but also to guide for adaptations, risks, and benefits. The use of educational strategies, with organized and systematized information before bariatric surgery, allows stimulating an attitude of self-care and lifestyle changes [4]. Empowering patients with this knowledge, allows them to realize that weight loss and lifestyle changes need to be maintained in the long term [5]. Working on these changes' pre-surgery increases the likelihood of a successful intervention [6].

In the follow-up of patients with chronic diseases and other comorbidities, the "case manager" nurses or nurse-led case-management interventions, improved the health of these patients, with improvement of health indicators. The figure of these specialized nurses induces quality of care and reduces health costs [7][8].

Good results of bariatric surgery depend on how patients experience, understand and accept the changes that develop in the postoperative period. It is in this context that nurse-led case-management may allow better management of the surgical process, through health education and promotion of healthy lifestyles, supported by a multidisciplinary team. These interventions have the potential to bring very positive results in self-reported quality of life, mental health, and reduction of cardiometabolic risk [9]. Thus, we intend to verify how a new eHealth nurse-led intervention can contribute to the quality of life, adaptation process and maintenance of weight loss of post bariatric surgery patients.

This paper's aim is to present the methodology of a research project that will design and evaluate this nurse-led intervention. After the Introduction in section I the methodology is described in section II, with an emphasis on the study design and the flow of the evaluation stage.

## II. METHODS

This research project will follow a mixed method approach, supported by a cycle of Design Science Research Methodology (DSRM) [10], that will allow us to design and evaluate a new intervention of specialized "nurse-led case-management" (Table 1).

TABLE I. DESIGN SCIENCE RESEARCH METHODOLOGY CYCLE

| ***DSRM Activity*** | **DESIGN SCIENCE RESEARCH METHODOLOGY CYCLE** | |
|---|---|---|
| | ***Objectives*** | ***Method/Tasks*** |
| Activity 1 | Identification of the problem and motivation | Systematic review of the literature |
| | | Know the context under study |
| Activity 2 | Set objectives for the intervention | Focus Group - Brainstorming Method |
| Activity 3 | Design and development of the NURLIFE program | E-health nurse-led case management program |
| Activity 4 | Demonstration of NURLIFE intervention | Short-term control case study |
| Activity 5 | Evaluation the results of the NURLIFE | RCT study, with intervention groups and control group, lasting approximately 14 months |
| Activity 6 | Communication of results | Communicate results from all development points |

### A. *Study design*

Activity 1- A systematic review of the literature with meta-analysis will be carried out with the objective of identifying interventions directed at bariatric surgery patients and the respective outcomes. There will also be interviews with patients who underwent surgery, as well as with professionals who are part of the current follow-up protocol, to understand the context where the intervention is developed and uncover possible points for improvement.

Activity 2 - Conduct two to three focus groups with representatives of the various stakeholders (patients and health professionals) in order to identify what the objectives for a new intervention should be. These objectives are expected to be related to the set of endpoints interest that will be used to measure the results of the intervention in the evaluation stage.

Activity 3 - In this activity, a nurse-led case-management intervention program is expected to be developed (NURLIFE). The aim of this intervention is to improve the management of the bariatric surgery process by patients in collaboration with the health care team, with a view to improving health and adoption of healthy lifestyles that enhance better outcomes. During this stage, the initial protocol of interprofessional collaboration, the skill mix of the multidisciplinary team, and the requirements of the web-platform where most of the communication with patients will occur, will be set.

Activity 4 - Conducting a short-term case-control study. The objective of this study will be to test the intervention designed in the previous step with a reduced number of participants, to optimize the intervention.

Activity 5 - Evaluating the results of the NURLIFE program. To proceed with this activity, a randomized clinical trial (RCT) will be conducted, with division of the participants into two groups (Figure 1).

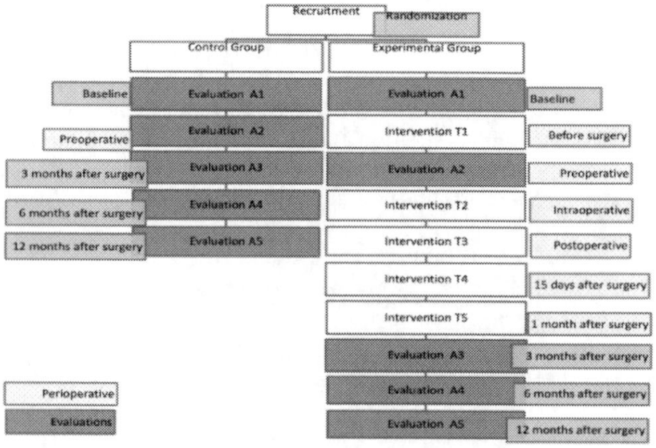


Figure 1. Prediction of the evolution and recruitment plan of the RCT study to evaluate the intervention

The predicted process and timeline for the RCT will be developed in all perioperative period.

### B. *Sample*

The sample size is based on the number of patients who had bariatric surgery in previous years, with the reference value of 100 patients per year, the usual in the context of the hospital. As inclusion criteria in the sample, participants will have to be enrolled for bariatric surgery at the Hospital of Évora, Portugal and agree to participate in the study. The main non-inclusion criteria will be patients with secondary bariatric surgery, with severe psychiatric or neurologic disease, and patients who do not agree to participate.

### C. *Randomization*

Each participant will be randomly assigned to each group after signing the informed consent and conducting the initial assessments. All data collected will be de-identified with identification ID, safeguarding confidentiality of the collected data.

### D. *Outcomes*

The primary endpoint of the RCT will be the proportion of patients who maintain their primary weight loss at the end of the first year.

For the measurement of clinical data, anthropometric parameters, surgical data, and a health data questionnaire will be used. The remaining variables will be defined based on activity 2 of the DSRM cycle, and may include life quality, physical activity, sedentary behavior, comorbidities.

## III. CONCLUSION

This project aims to be the first study to investigate the effect of specialized interventions on patients who are candidates for bariatric surgery. It is expected that this project will provide evidence of the impact ofevidencebased mixed, face-to-face and e-health programs, on the long term maintenance of bariatric surgery results.

## REFERENCES

[1] K. J. Foreman, N. Marquez, A. Dolgert, K. Fukutaki, N. Fullman, M. McGaughey, et al., "Forecasting life expectancy, years of life lost, and all-cause and cause-specific mortality for 250 causes of death: reference and alternative scenarios for 2016–40 for 195 countries and territories" The Lancet, pp. 2052–90, 2018.

[2] J. L. Colquitt, K. Pickett, E. Loveman and G. K. Frampton, "Surgery for weight loss in adults" Cochrane Database Syst Rev. 2014, 2014

[3] J. I. Mechanick, A. Youdim, D. B. Jones, W. T. Garvey, D. L. Hurley, M. McMahon, et al., "Clinical Practice Guidelines for the Perioperative Nutritional, Metabolic, and Nonsurgical Support of the Bariatric Surgery Patient. The Obesity Society, and American Society for Metabolic & Bariatric Surgery" Obes Silver Spring Md, 2013

[4] L. M. Barros, F. N. Carneiro, N. M. Galindo-Neto, M. F. Araújo, R. A. Moreira, L. P. Barbosa, et al., "Educational intervention and obesity indicators in gastroplasty candidates: quasi-experimental study" Acta Paul Enferm, 2020.

[5] A. Petcu, "Comprehensive Care for Bariatric Surgery Patients" AACN Adv Crit Care, 28(3):263–74, 2017

[6] E. Yildiz and S. Karagözoğlu, "The Effects of Nursing Education Constructed According to Roy Adaptation Model on Adaptation Process of Patients Undergoing Bariatric Surgery. Bariatr Surg Pract Patient Care" 98–108, 2021

[7] V. Doménech-Briz, M. R. Gómez Romero, I. Miguel-Montoya, R. Juárez-Vela, J. R. Martínez-Riera, M. Mármol López, et al, "Results of Nurse Case Management in Primary Heath Care: Bibliographic Review" Int J Environ Res Public Health, 2020.

[8] S. K. Chow and F. K. Wong, "A randomized controlled trial of a nurse-led case management programme for hospital-discharged older adults with co-morbidities" J Adv Nurs.2257–71, 2014

[9] G. Cangelosi, I. Grappasonni, P. Pantanetti, S. Scuri, G. Garda, N. Cuc Thi Thu, et al., "Nurse Case Manager Lifestyle Medicine [NCMLM] in the Type Two Diabetes patient concerning post COVID-19 Pandemic management: Integrated-Scoping literature review" Ann Ig 2022 Feb 8. doi: 10.7416/ai.2022.2500.

[10] L. V. Lapão, M. M. Da Silva, and J. Gregório, "Implementing an online pharmaceutical service using design science research" BMC Med. Inform. Decis. Mak., vol. 17, no. 1, 2017, doi: 10.1186/s12911-017-0428-2

# Electronic Health Records User Experiences: a Nationwide Survey From Norwegian Hospitals.

[a] Ove K Lintvedt, [b] Maryam Tayefi, [c] Espen Nordheim, [d] Rune Pedersen, [e] Kristian Malm Nicolaisen, [f] Halvard Lærum, [g] Bente S. Nedrebø, [h] Luis Marco-Ruiz

Norwegian Centre for E-health Research, University Hospital of North Norway, Tromsø, Norway

Department of Architecture Governance, The Norwegian Directorate of eHealth, Oslo, Norway

Research and Knowledge Resources, Sikt, Bergen, Norway

e-mail: [a]ove.lintvedt@ehealthresearch.no, [b]maryam.tayefi@ehealthresearch.no, [c]espen.solbakken.nordheim@ehealthresearch.no, [d]rune.pedersen@ehealthresearch.no, [e]kristian.nicolaisen@ehealthresearch.no, [f]hallvard.laerum@ehelse.no, [g]bente.nedrebo@sikt.no, [h]luis.marco.ruiz@ehealthresearch.no

***Abstract*—The adoption of a new Electronic Health Record (EHR) is a disruptive event for hospitals influencing the satisfaction and performance of clinicians. In Norway, the four health regions (South-East, West, Central, and North) have used several EHR systems for decades. This study analyses the satisfaction of clinical users and determines which features of the EHR that should be prioritized to improve clinicians' satisfaction. In addition, findings show a relatively high frequency of interruptions that could affect secure and easy access for the health care professionals to information about the patients. Finally, differences are found within the same EHR regarding clinical user satisfaction, meaning that the context significantly impacts the satisfaction.**

***Keywords-Electronic Health Records; Usability; User Satisfaction; Human Factors; Computerized Clinical Decision Support Systems (CCDSS).***

## I. INTRODUCTION

Digitalization programs in the EU and US have spread the adoption of Electronic Health Records (EHR) [1][2]. EHRs have evolved from electronic journals for note keeping into integrated health information systems that provide a holistic view of most patient clinical information and actively support clinical users in decision making through Computerized Clinical Decision Support Systems. Examples include infobuttons, medication alerts, and computerized medication order entry [3]. While the functionalities of EHRs have been significantly expanded, there are concerns about their actual usefulness and perceived benefit [4]. Some studies point out that the amount of work required by EHR interaction contributes to clinical burnout due to a focus on long-term outcome measures and reimbursement rather than the actual value of daily practice [5]. International research is often focusing on usability, safety, quality, functionality, and satisfaction [6][7][8]. Developing e-health indicators is a relatively new focus in the field. Nordic eHealth Research Network (NeRN) works on standard Nordic e-health indicators [9].

In the last decade, Norway has increased the investments on eHealth initiatives to meet the future need of the healthcare sector. This wave of digitalization is in line with the national white paper 9, 'One citizen – one Health Record' [10]. This white paper states that: a) secure and easy access for the health care professionals to information about the patients is required; b) citizens should have access to safe and accessible digital services; and, c) clinical data should be available for monitoring, management, and research.

One of the actions towards addressing these requirements has been the deployment of a newer EHR system in three (North, South-East, and West) out of four health regions. A fact that directly impacts clinicians' daily work.

This change has raised the need to evaluate clinicians' user experience to understand if the investments in new EHRs have improved clinical users' experience. A previous study in 2019 published an initial evaluation with three hospitals from two health regions and a total of 208 wholly answered questionaries [11]. This study targeted the use of the currently implemented EHR systems, focusing on clinical task support and the overall satisfaction with the EHR. Each of these three hospitals had implemented DIPS (Distributed information and patient data system in hospitals) Classic (DIPS, DIPS ASA, Bodø) over the previous years, 2010-2014. However, these three hospitals covered only two main cities in Norway, corresponding to two neighbouring regional health administrations (South-East Norway and West Norway, respectively). This paper expands the former study by including data from University Hospital North Norway (North Norway health region) and St. Olav Hospital (Central Norway health region), which allows us to cover all the four health regions in Norway.

Nurses and doctors were asked to specify functionalities that worked well in their EHRs and functionalities that required further improvement. This extension of the 2019 study attempts to determine if the results from the previous survey also hold for all Health Regions in Norway and elucidate the effect of the latest eHealth developments on clinical users.

This paper is organized as follows: Section II explains the methods used, where questionnaires and statistical methods are explained. Section III presents the survey results, focusing on the three types of satisfaction. Tables and figures illustrate the results. Section IV describes ethical considerations. In section V, general findings of the three types of satisfaction are discussed, and a comparison between national data and data from the study including only two health regions. Section VI concludes in relation to significant findings from this national survey. In addition, it summarises how the findings can be understood and how suppliers and regional health authorities can make use of the findings.

## II. Methods

In this section we focus on the questionnaire, statistical methods in use, as well as the data collection.

### A. *Setting*

We approached clinical users from the newly implemented EHRs at Haukeland University Hospital (HUH), University Hospital of North Norway (UNN), Trondheim University Hospital (St. Olav), and Oslo University Hospital (OUH). These hospitals belong to the four Regional Health Authorities in Norway, which allows us to cover the whole country. Except for St. Olav, all hospitals had been involved in the transition from their previous system, DocuLive (DocuLive EPR, Siemens Medical Systems Norway, Oslo)., into the new system, DIPS Classic. Specifically, OUH, HUH, and UNN have been implementing the new her since 2010, and St. Olavs has been implementing theirs since 1999.

Before 2006, DocuLive was a system of journal documents mainly used by clinicians for reading and signing clinical documents and laboratory results. The system lacked integration with the patient administrative system. DocuLive added functionality to scan documents into the system, e.g., forms and response reports. At the same time, DIPS Classic was regarded as a complete system with a patient administrative system, patient journal, laboratory system, and an integrated system for psychiatry [12].

The fundamental reasons for changing the system were a lack of necessary functionality and a lack of integration between the journal system and the patient administrative system. In addition, the government took over the ownership of the public hospitals in Norway in 2002 [13]. The regional health trusts promoted the standardisation of health information systems. For example, in the Northern Health Region, where UNN was the only hospital in the health region using DocuLive, switching to DIPS was seen as a common-sense decision.

### B. *Data collection*

A total of 506 clinicians (nurses and physicians) were contacted in 2018. Response rates were 35.0%, 22.0%, and 29.0% for physicians, nurses, and all clinicians, respectively. Surveys from UNN, St. Olav, and HUH were gathered with the following responses: n=87 physicians, n=60 nurses, n=147 in total. We issued ten reminders to the respondents in the period from September to December 2018 and four reminders to their superiors. Some superiors also reminded the respondents in person at a joint meeting. The availability of clinical functionality varied between units and wards in the same units.

Email lists for the employees in the units were collected from the institutions themselves. Respondents were selected using a random number generator [14].

For OUH, surveys were gathered from 2015/2016 (n=152 physicians), and details about data collection, methods, and results are published elsewhere [11].

Regarding inclusion criteria, participants were physicians and nurses who worked full time at any participant hospitals in 2018, and only physicians who worked full time in 2015/16. Respondents who stated that they did not actively work with patients were excluded. A total of 299 individuals, of which 239 (79.9%) were physicians and 60 (20.1%) were nurses, completed the questionnaire (see Table 1). Forty-six worked at Haukeland University Hospital (HUH), 39 worked at University Hospital of North Norway (UNN), and 62 worked at St. Olavs hospital – Trondheim University Hospital (St. Olav), and 152 worked at OUH. Data from OUH was only available for one year (2015/16), but no new questionnaires were filled out for the following years in this hospital. Still, we kept the results from OUH to cover the regional health area. Data were stratified by both time and organization to deal with this limitation. Table I shows the amounts of respondents to questionnaires gathered by hospital and clinical role.

### C. *Questionnaire*

The web-based survey tool Questback (Questback, Oslo, Norway) was used for the online questionnaire. Questback allows users to record their responses anonymously and preserve their privacy remotely. The questionnaire was piloted among nine representatives of the target group. Some content was revised based on feedback from the pilot group. Questions were dynamically designed to ensure that the respondents were only asked about relevant things for their stated place of work and role (physician or nurse). Some questions were sent to both nurses and physicians, while parts of the survey specifically targeted one of the professions. The survey also included several broad questions, which the respondents were allowed to answer freely.

The survey was developed based on past research using a previously validated questionnaire [15][16]. Changes were made as the old questionnaire was too extensive and tailored for physicians only. This new questionnaire is an early effort toward developing clinical user satisfaction and interruptions indicators. The new questionnaire used a 5-point Likert scale ('Completely disagree', 'Partially disagree', 'Neutral', 'Partially agree, 'Completely agree'). Some questions were only rated as agree/disagree. The overview of the questions in the questionnaire can be accessed online [17].

The questionnaire was structured in 3 main sections, all related to various dimensions of satisfaction. The first dimension was satisfaction with the EHR functionalities, which contained 11 items. The second dimension was related to generic aspects of the EHR system, which contained four items. The third dimension was related to overall satisfaction with the EHR system, which contained only one item.

### D. *Analysis/statistical methods*

The main statistical methods used for analysis were frequency (percentage) for discrete variables and mean for continuous variables. The Pearson chi-square test was used for comparison. The level of significance was considered as 0.05.

The statistical software SPSS 25 version was used for the analysis. In the process of cleaning data, we had to address missing values. For EHR functionality satisfaction, there were no missing values. For EHR generic satisfaction and EHR overall satisfaction, missing values were n=48 (15.7%) and n=47 (14.6%), respectively. The easiest way is to remove the incomplete records. In the case of lacking enough sample size, this procedure is not proper to deal with missing values. Several imputation techniques have been suggested in situations where the missingness is completely random (MCAR). MCAR implies no systematic reasons for missingness [18]. Little [19] provided the chi-square test for the MCAR assumption to check that the missingness is completely random. While providing the comparison between health regions/satisfaction profession/satisfaction, we applied the MCAR assumption by Little (Chi-Square=502.988, df=512, p=.604). The results confirmed that the missingness is MCAR. Then, we imputed our missing records with the neither-nor value.

### E. *Ethics*

The Regional Committee for Medical and Health Research Ethics South-East Norway has been consulted. According to national regulations ethics, approval was not required because the study did not involve biomedical research, and all data were anonymized.

## III. Ease of Use

This section will present the baseline data and questionnaire results. Further there will be a presentation of the three types of satisfaction.

### A. *Baseline data*

Table I contains the baseline data by year, location, and participants clinical role. The medical field with the highest number of participants was the aggregation of those treating surgical, woman-related, and cancer conditions, with n=138 (46.0%). The following fields with the highest number of participants were related to neurological, orthopaedics, and rehabilitation conditions n=98 (33.0%), followed by medical, heart/lung, and other conditions with n=35 (12.0%), 21 (7.0%), and 7 (2.0%), respectively.

### B. *Questionnaire results*

There were two questions related to interruptions of the clinical workflow while using the EHR. The first one regarded interruption caused by login requests; results range from 4 to 50 interruptions per day (outliers removed). The mean number of interruptions per day is 17.21. The corresponding number from the 2019 study was 17.15. The second one regarded the number of interruptions due to the EHR hanging or crashing. The mean number of interruptions is 3.08, corresponding to one interruption per week in the scale used. The corresponding numbers from the 2019 study was 2.95, slightly less than once a week. Almost 70% of respondents reported interruptions that ranged between once a year and once a week, 30% that interruptions occurred once or more a day. The corresponding numbers from the 2019 study was 72% and 28%, respectively.

TABLE I. Baseline Dataset

| Health Region (Survey year) | Clinical profession | | |
|---|---|---|---|
| | *Physician* | *Nurse* | *Total* |
| West (2018) | 34 | 12 | 46 (15.4%) |
| Central (2018) | 31 | 31 | 62 (20.7%) |
| North (2018) | 22 | 17 | 39 (13.1%) |
| South-East (2016) | 152 | 0 | 152 (50.8%) |
| Total | 239 (79.9%) | 60 (20.1%) | 299 (100.0%) |

### C. *EHR functionality satisfaction*

Table II presents the responses about satisfaction with specific EHR functionalities.

Regarding aggregated EHR function satisfaction for all 11 questions, 47.0% of respondents reported to be satisfied, 35.0% were neither satisfied nor dissatisfied, and 18.0% were not satisfied (see Fig. 1). The corresponding numbers from the 2019 study was 59.0%, 26.0%, and 15.0%, respectively.

If we observe the functionalities in Table II, the one rated with the highest satisfaction was question 1, where 72.0% of respondents were satisfied with the function allowing them to read sample responses from medical biochemistry. The lowest satisfaction was related to question 11, which refers to the overall overview of the patient´s drug treatment, where only 29.0% of the respondents were satisfied. The corresponding numbers from the 2019 study was 84.0% and 33.0%, respectively.

Questions 2 to 5 have the second-highest satisfaction rate, between 55.5% to 69.0%. For these questions, the range of respondents who answered that they were neither satisfied nor dissatisfied was 11.0% to 17.5%. The corresponding numbers from the 2019 study was between 58.0% to 70.0% and 19.0% to 31.0%, for satisfaction and neither-nor, respectively.

TABLE II. EHR Functionality Sasisfaction

| Survey question | EHR Functionality Satisfaction | | |
|---|---|---|---|
| | *Satisfied* | *Neither- nor* | *NotSatisfied* |
| 1 Read sample responses from medical biochemistry | 72.5% | 21.0% | 6.5% |
| 2 Compare the treatment and efficacy of a particular patient | 69.0% | 20.0% | 11.0% |
| 3 Overview of the patient´s issues | 60.0% | 27.0% | 13.0% |
| 4 Read the radiology response reports | 55.5% | 27.0% | 17.5% |
| 5 Overview of your outstanding task | 55.5% | 33.0% | 11.5% |
| 6 Communicate with patient about health information | 25.5% | 51.5% | 23.0% |
| 7 Receive specific advice and recommendations for further treatment | 33.0% | 51.0% | 15.0% |
| 8 Prescribe drug treatment | 33.0% | 44.0% | 22.0% |
| 9 Concrete plan for the patient´s assessment, treatment and care | 39.5% | 33.0% | 11.5% |
| 10 Assess the right to priority health care | 32.0% | 37.0% | 31.0% |
| 11 Overall overview of the patient´s drug treatment | 29.0% | 31.0% | 34.0% |

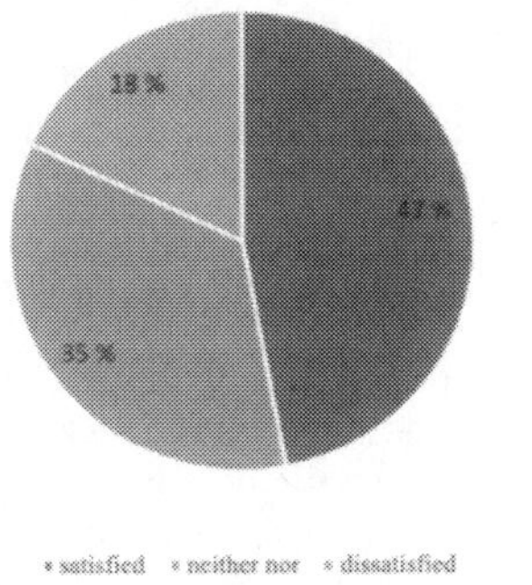


Figure 1. EHR functionality satisfaction.

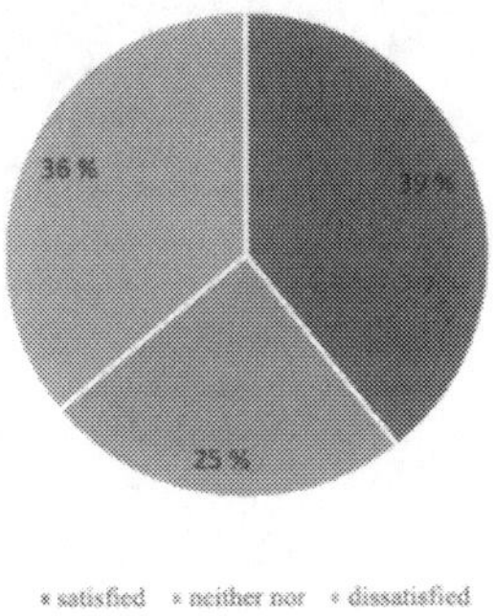


Figure 2. EHR overall satisfaction.

Questions 11, 10, 6, and 8 have the highest dissatisfaction rates, with more than a 20.0% of respondents answering that they were not satisfied. This is the same as for the 2019 study.

Questions 6, 7, and 8 have the highest indifferent rate (nighter satisfied nor dissatisfied). From the 2019 study, questions 11, 9, 6 ans 7 have the highest indifferent rate.

For the aggregated dataset containing all health regions, the differences by clinical role differences were not significant ($p<0.603$). Nurses (20.4% satisfaction) are less satisfied than physicians (79.6%), which is not significant ($p>0.05$).

There were no significant differences among health regions ($p>0.05$). Likewise, the difference between the central region (St. Olav), which used the DocuLive system, and the other three regions, which used DIPS Classic, was not significant ($p>0.05$).

### *D. EHR generic satisfaction*

Generic EHR satisfaction refers to effectiveness, high quality, worth time and effort, and user-friendliness.

A total of 39.3% of respondents were satisfied, 24.8% of respondents reported that they were neither satisfied nor dissatisfied, and the remaining 35.9% reported being dissatisfied (see Fig. 2). The corresponding numbers from the 2019 study was 40.1%, 23.2%, and 36.7%, respectively.

By clinical role, the difference was significant ($p<0.001$). Nurses (33.6% satisfaction) are less satisfied than physicians (66.4% satisfaction).

The generic satisfaction between regions West (18.0% satisfaction), Central (25% satisfaction), North (18.8% satisfaction), and South-East (38.3% satisfaction) was significant ($p<0.001$). However, when comparing the region with DocuLive (Central) with the other three regions using DIPS Classic, no significant differences were found for EHR generic satisfaction ($p>0.05$).

### *E. EHR Overall satisfaction*

Overall satisfaction was addressed through a single item. A total of 34.7% of respondents were satisfied, 24.4% of respondents reported that they were neither satisfied nor dissatisfied, and the remaining 40.8% reported being dissatisfied (see Fig. 3). The corresponding numbers from the 2019 study for overall satisfaction was 37.5%, 21.5%, and 41.0%, respectively.

The overall satisfaction by clinical role was significant ($p<0.001$). Nurses (36.3% satisfied) are less satisfied than physicians (63.7% satisfied).

The overall satisfaction by regions West (17.6%), Central (22.0%), North (20.9%), South-East (39.6%) was found to be significant ($p<0.05$). However, the overall satisfaction for the region with DocuLive (Central) was not significant when compared with the group of the other three regions using DIPS Classic ($p> 0.05$).

## IV. DISCUSSION

The discussion will focus on different types of satisfaction and the more generic findings.

### *A. EHR functionality satisfaction*

Four questions had the lowest score about EHR functionality satisfaction. These were related to the overall overview of patients' drug treatment, the assessment of the proper priority of care, the concrete plan for patient assessment, communication with the patient about health information, and prescriptions of drug treatments. These are the main functionalities that vendors should prioritize when developing new versions of their products.

### *B. EHR generic satisfaction*

Most of the respondents were satisfied with the generic functionality of the EHR. The proportion of respondents being indifferent or dissatisfied adds up to almost 61%. Although this does not indicate a clear need for product replacement, it

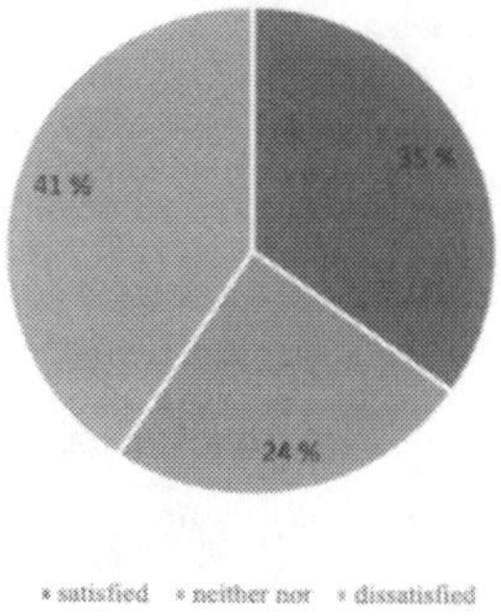


Figure 3. EHR generic satisfaction.

points out a significant room for improvement in current EHRs' generic satisfaction. Vendors should prioritize understanding why almost 36% of users are dissatisfied. Also, they should understand more about the requirements from users in general to improve satisfaction, so more users are fully satisfied.

### C. *EHR overall satisfaction*

With regards to overall satisfaction, there is a significant portion of respondents that reported to be overall dissatisfied (40.8%). Interestingly, this level of dissatisfaction is higher than the score found when measuring specific functionality dissatisfaction. This may indicate that, even though users were not dissatisfied with specific functionalities, the integration of functionalities in the system workflow increases overall dissatisfaction. Vendors should consider better integration of their functionality in clinicians' workflow.

### D. *Generic findings*

Differences in overall satisfaction were significant depending on the clinical role and could be interpreted as differences in the satisfaction of role-specific functionality. Physicians were overall more satisfied than nurses for the three dimensions of satisfaction and could be interpreted as a requirement for improving nurse-specific functionality.

No significant differences were found in the EHR used (DocuLive vs. DIPS classic). However, there were significant differences among regions regarding generic satisfaction and overall satisfaction. We interpret this finding as a difference arising from differences in the deployment context, but there is no evidence of this difference coming from the system itself (DIPS Classic vs. DocuLive).

The high number of interruptions (number of logins a day and EHR hanging or crashing) could directly affect security and easy access for healthcare personnel. The high frequency of interruptions indicates that the first goal from the government stated in the whitepaper "one citizen - one Health Record" (secure and easy access for the health care professionals to information about the patients is required) is still not covered good enough. Missing on this goal will have consequences for both citizens and healthcare professionals. It should provide patients and residents with safer and better treatment and health professionals a more uncomplicated working day.

### E. *Comparison with study including two health regions*

The previous study from 2019 [11] included only physicians from two health regions. For this reason, there will be no comparison of regions or professions.

There are no significant differences in the number of interruptions of the clinical workflow while using the EHR.

Aggregated EHR function satisfaction for all 11 questions shows almost the same response for dissatisfaction. More respondents were satisfied in the 2019 study. Both studies report the same function with the highest satisfaction and the same for the function with the lowest satisfaction for the EHR functionalities. The response pattern for the other functionalities is almost similar between the two studies. Our interpretation is that the decrease in satisfaction could be a regional effect as OUH got the new DIPS Classic with new functionality one year before the survey. The new regions had several years of experience with the system when the responded to the survey. The inclusion of nurses could cause one other explanation as they reported less satisfaction in this new study.

Generic EHR satisfaction is almost identical for the two studies. It seems like the inclusion of more hospitals/regions, and other professions do not change this effect. Even though, the EHR functional satisfaction is lower in the new study, the satisfaction regarding effectiveness, perceived quality, and user-friendliness is persistent.

The EHR overall satisfaction was addressed through a single item and persistent on a national level.

## V. CONCLUSION

This study has analysed hospitals from the four health regions in Norway. No differences in generic or overall satisfaction were found when including data for all four health regions in Norway. Data from the study using only two health regions was higher than for the one with all four health regions. No differences in satisfaction were found between the two types of EHRs. Significant differences are found within the same EHR, meaning that the context significantly impacts the satisfaction attributed to the EHR. A substantial portion of users reported not being fully satisfied with their EHR. Nurses were the group less satisfied, and functionalities related to showing the broad overview of patients are where vendors should concentrate their efforts to respond to user requirements and improve their satisfaction.

The EHR vendors and regional health authorities should also be focusing on bridging the gap between the goal for secure and easy access to patient data and the relatively high number of interruptions experienced by the clinical users.

## ACKNOWLEDGMENT

This study was funded by the Norwegian Centre for E-health Research.

## REFERENCES

[1] European Commission, Directorate-General for the Information Society and Media, "Accelerating the development of the eHealth market in Europe". Publications Office, 2008. [Online]. Available from: https://data.europa.eu/doi/10.2759/19946, accessed Mar. 22, 2022.

[2] D. Blumenthal and M. Tavenner, "The 'Meaningful Use' Regulation for Electronic Health Records," N. Engl. J. Med., vol. 363, no. 6, pp. 501–504, Jul. 2010, doi: 10.1056/NEJMp1006114.

[3] R. A. Greenes, Clinical Decision Support: The Road to Broad Adoption. Academic Press, 2014.

[4] A. D. Black, J. Car, C. Pagliari, C. Anandan, K. Cresswell, T. Bokun, et al., "The impact of eHealth on the quality and safety of health care: a systematic overview," PLoS Med., vol. 8, no 1, Jan. 2011, doi: 10.1371/journal.pmed.1000387

[5] C. P. West, L. N. Dyrbye, and T. D. Shanafelt, "Physician burnout: contributors, consequences and solutions," J. Intern. Med., vol. 283, no. 6, pp. 516–529, Jun. 2018, doi: 10.1111/joim.12752.

[6] M. Zahabi, D. B. Kaber, and M. Swangnetr, "Usability and Safety in Electronic Medical Records Interface Design: A Review of Recent Literature and Guideline Formulation." Hum Factors. vol. 57(5), pp. 805-34, Aug. 2015, doi: 10.1177/0018720815576827.

[7] A. D. Black, J. Car, C. Pagliari, C. Anandan, K. Cresswell, T. Bokun, et al. "The impact of eHealth on the quality and safety of health care: a systematic overview", PLoS Med., vol. 8(1), jan. 2011, doi: 10.1371/journal.pmed.1000387

[8] J. Kaipio, T. Lääver, H. Hyppönen, S. Vainiomäki, J. Reponen, A. Kushniruk, et al. "Usability problems do not heal by themselves: National survey on physicians' experiences with EHRs in Finland." Int J Med Inform. Vol. 97, pp. 266-281, Jan. 2017, doi: 10.1016/j.ijmedinf.2016.10.010

[9] C. Nøhr, A. Faxvaag, C. H. Tsai, G. A. Harðardóttir, H Hyppönen, H. K. Andreassen, et al., "Nordic eHealth Benchmarking. Towards evidence informed policies," 2005, doi: 10.6027/temanord2020-505

[10] Ministry of Health and Care Services, "Whitepaper. no. 9 (2012-2013): 'One citizen – one Health Record'. St.Meld. nr. 9 (2012–2013) 'Én innbygger - én journal'. Nov. 27, 2012. [Online]. Available from: https://www.regjeringen.no/no/dokumenter/meld-st-9-20122013/id708609/ , accessed Mar. 22, 2022.

[11] T. R. Schopf, B. Nedrebø, K. O. Hufthammer, I. K. Daphu, and H. Lærum, "How well is the electronic health record supporting the clinical tasks of hospital physicians? A survey of physicians at three Norwegian hospitals," BMC Health Serv. Res., vol. 19, no. 1, p. 934, Dec. 2019, doi: 10.1186/s12913-019-4763-0.

[12] G. Ellingsen, and E. Monteiro. "The slight surprise of integration", in: C. Sørensen, Y. Yoo, K. Lyytinen, J. L. DeGross (eds), Designing Ubiquitous Information Environments: Socio-Technical Issues and Challenges. IFIP — The International Federation for Information Processing, vol 185. Springer, Boston, MA, pp.261–274, 2005, doi: https://doi.org/10.1007/0-387-28918-6_20

[13] A. Hysing. "Dips også i Helse Nord," Computerworld, June 2014, [Online]. Available from: https://www.cw.no/dips-ogsa-i-helse-nord/935317 , accessed Mar. 22, 2022.

[14] https://www.random.org/

[15] V. Heimly, A. Grimsmo, T. P. Henningsen, and A. Faxvaag, "Diffusion and use of Electronic Health Record Systems in Norway," MEDINFO 2010, pp. 381–385, 2010, doi: 10.3233/978-1-60750-588-4-381.

[16] H. Lærum and A. Faxvaag, "Task-oriented evaluation of electronic medical records systems: development and validation of a questionnaire for physicians," BMC Med. Inform. Decis. Mak., vol. 4, no. 1, p. 1, Feb. 2004, doi: 10.1186/1472-6947-4-1.

[17] Online overview questionnaire (https://www.researchgate.net/publication/359426633_survey_2018#fullTextFileContent).

[18] R. J. A. Little and B. Rubin, "The analysis of social science data with missing values," Sociological Methods and Research, vol. 18, no 2-3, pp. 292-326, Nov. 1989, doi: 10.1177/2F0049124189018002004

[19] R. J. A. Little, "A test of Missing Completely at Random for multivariate data with missing values," Journal of the American Statistical Association, vol. 83, pp. 1198-120, Des. 1988, doi: 10.2307/2290157

# Intended and Unintended Consequences of Implementing a Nursing App

Gro-Hilde Severinsen
Norwegian centre for E-health research,
University hospital North Norway,
Tromsø, Norway
Email: Gro-Hilde.Severinsen@ehealthresearch.no

Line Silsand
Norwegian centre for E-health research
University hospital North Norway
Tromsø, Norway
Email: Line.Silsand@ehealthresearch.no

Kristian Malm-Nicolaisen
Norwegian centre for E-healht research
University hospital North Norway
Tromsø,Norway
Email: Kristian.Nicolaisen@ehealthresearch.no

Rune Pedersen
Norwegian centre for E-health research,
University hospital North Norway
[2]UiT Arctic university Tromsø
Tromsø, Norway
Email: Rune.Pedersen@ehealthresearch.no

Beate Sørslett
Nordlandssykehuset
Bodø, Norway
Email: Beate.Sorslett@nlsh.no

Gunnar Ellingsen
Norwegian centre for E-health research,
University hospital North Norway
[2]UiT Arctic University Tromsø
Tromsø, Norway
Email: Gunnar.Ellingsen@uit.no

***Abstract* - The paper addresses socio-technical consequences from implementing a nursing app in a North Norwegian health trust. We used the NASSS-CAT (Non-adoption or abandonment of technology by individuals and difficulties achieving Scale-up, spread and sustainability) framework for evaluating the app implementation. The paper is relevant to the Nurse team application track, presenting nurses experience with adoption and use of a nursing app for monitoring progression or deterioration trough measuring vital signs. In this paper, the focus is on emerging downstream values for both the healthcare workers and the organization, since these factors are well suited to indicate whether the implementation is a success or not. From the analysis, both intended and unintended consequences of the implementation are outlined. The unintended consequences illustrated how extensive repercussions implementing a simple nursing app can have for the workflow of both nurses, doctors and leaders, as well as the communication at different healthcare levels. In addition, implementing an app for clinical practice generated improvement in the digital competence for both healthcare personnel and the organization.**

***Keywords - Intende consequences; unintended consequences; nursing app; implementation; qualitative evaluation.***

## I. Introduction

Documentation of and actions taken related to patients' vital signs are of fundamental importance for clinical outcomes [1]. The paper is relevant to the Nurse team application track since it presents nurses experience with adoption and use of a nursing app for monitoring progression or deterioration trough measuring vital signs.

In most Northern European countries, for instance in Norway and the UK, monitoring patients' clinical status is guided by protocols based on Early Warning Scores (NEWS). NEWS includes vital signs like blood pressure, heart rate, respiration rate and oxygen saturation [1][2]. These are key indicators of a patient's physiological status. Based on the vital parameters, NEWS allows clinicians to produce a score ranging from 0 to 20, where a higher score indicates greater clinical risk. A deterioration in a patient's condition is often detected by abnormalities in vital signs, and failure to detect deterioration at an early stage is associated with worsened patient outcomes and a contributing factor for avoidable deaths [1][3]. Therefore, nurses measure vital signs regularly, especially for acutely unwell patients [2][3]. In general, patients with a low NEWS score are monitored every 6-12 hours, increasing to hourly for patients with a score above 6 [4]. This results in healthcare professionals conducting an extensive number of time-consuming NEWS scores [5][6]. Nurses often report challenges with maintaining essential patient surveillance due to high workload e.g., one study found that around 35% of the vital sign assessments scheduled according to an early warning score-based protocol were delayed or missed.

Today, measuring the vital signs is mainly a paper-based procedure where the score is registered on paper, and the paper forms are scanned into the Electronic Health Record (EHR) system. However, even if NEWS is implemented as part of clinical practice, previous research has shown that transferring the information from one medium to another has been challenging, in addition, nurses often write vital signs onto interim paper notes prior to documenting them in the EHR, resulting in double registration practices. Transferring information implies not only time delays, but also an increased number of transcription errors, which could lead to inappropriate treatment and affect patient outcome [2]. It is important that the early warning score is evaluated in line with the overall condition of the patient [2], and that nurses communicate vital signs information as fast and efficiently as possible. Removing the barriers to more direct vital sign documentation can reduce these challenges and facilitate more accurate and timely documentation procedures, contributing to safer patient care [1][5]. Understanding todays nursing workflows and their' view of clinical and documentation processes can help optimize recording of point of-care vital signs documentation. Hence, electronic support tools facilitating timelier and safer recording of vital signs at the point-of-care are important to explore [1][2]. However, the mechanisms involved in implementing and using digital tools for NEWS registration is mainly unvisited both in the healthcare practice and the literature

To that end, the aim of the paper is to present and discuss empirical findings from the implementation of a nursing app. The app is used for registering vital signs and calculating NEWS score at the point-of-care in a Health Trust (HT1) in Norway. In this paper, we outline consequences generated by the implementation. Hence our research question is: *What are the intended and unintended socio technical consequences of implementing a nursing app?*

Our empirical site is HT1in the North Norwegian health region. Employing 4000 healthcare professionals, HT1 encompasses 4 hospitals and provide services for 136000 patients [7]. In 2020, HT1 implemented an app for registering vital signs and NEWS score at the patient's bedside. The overall aim of the implementation was to improve the nurses' workflow and patient safety, in addition to improving the digital competence of the users as well as the organization. There is a globally shared concern related to health personnel's digital competence, and skills required for adopting new digital services [8].

In the paper a qualitative evaluation approach is used mainly based on semi-structured interviews with different actors involved in the implementation and use of the app. The evaluation is analysed using the NASSS-CAT (Non-adoption or abandonment of technology by individuals and difficulties achieving Scale-up, spread and sustainability) framework designed by Trisha Greenhalg et al [9].

The rest of the paper is structured as follows: Section II describes the empirical project and the background for the implementation. Section III present our qualitative evaluation approach including the data collection and the NASSS-CAT framework focusing on the category value proposition to discuss the socio technical consequences of the implementation. In Section IV, the results of the study are presented, including the intended and unintended consequences of implementing the nursing app. In section V, the success of the app implementation is discussed, including the importance of digital maturity. Section VI outlines the conclusion and the implications from the study.

## II. Background

HT1 had priorly implemented a procedure for registering NEWS on all patients both in somatic and psychiatric care. This implied that the nurses had to observe and measure the patient's vital signs at the bedside and write the results of the measures on a paper note. Back in the ward office, they transferred the vital signs into a paper form and calculated the NEWS for each patient. The nurses then plotted every single score into the patients' paper-based Medical Charting Schema (MCS). Hence, calculating NEWS implied double and triple registration of vital signs. The procedure of registering NEWS was conducted every morning before the medical doctor's ward round, and the doctors had to go to the ward and search for the paper form to find updated NEWS scores for each patient. If a patient's condition was severe or worsened during the shift, NEWS was recorded more frequently to monitor the patient's situation closely. In addition, the nurses had to manually coordinate the work with NEWS score. This included checking with each other before conducting a NEWS score, to avoid measuring the vital signs twice because another nurse already had the results on a note in her pocket. Accordingly, it was a time-consuming process for the nurses, with an extensive risk of errors due to the complex transmissions of the data.

One of the health trusts in the North Norwegian health region (HT1), had for years been looking for a digital tool to improve the workflow and procedure of registering vital signs and calculating NEWS, in addition to an overall aim of removing paper forms and increasing data accessibility for both nurses and doctors. This required an app that was sufficiently user friendly and intuitive to use in a busy clinical setting. In addition, it was necessary that the app could be integrated with the existing IT infrastructure and automatically transfer data to the EHR system. It was important to find a solution that did not delay the work with complicated processes for logging in and registration of vital signs. The digital solution had to be more efficient than the existing paper-based routines to be of value for the organization.

HT1 was introduced for several vendors of digital NEWS score tools without finding anyone fit to fulfil their requirements. However, in 2019 HT1 was presented for a nursing app by a Finnish vendor, which had implemented their solution in several Finnish and Swedish hospitals as well as nursing homes. The Finnish vendor also had established a business collaboration with the EHR vendor in the North Norwegian health region. The app had the potential to integrate with several type of standards to exchange data with and reuse the data from the EHR system. However, at HT1 the EHR system was not able to receive or reuse standardized data elements, so the solution was to generate PDF files to be exported from the app to the EHR. There is a plan to include structured data in the EHR further on, then the app will be integrated more seamless with the system.

In short, HT1 found the app to comply with their needs, and came to an agreement with the Finnish Vendor and the hospital's EHR vendor. HT1 started the implementation of the app as a pilot project in an internal medicine department in March 2020. The pilot project was a success, and scaling of the implementation followed a snowballing effect where both medical, surgical, and mental health wards subsequently implemented the app. In autumn 2021, 26 wards at HT1 used the app, pertaining over 1000 users.

## III. METHOD

The paper is based on a qualitative inductive evaluation study conducted in close collaboration with HT1.

### A. *Data Collection*

In the study, 22 semi-structured interviews were conducted with healthcare personnel and other actors involved in the implementation and use of the app (see Table I)

TABLE I. OVERVIEW OF THE DATA COLLECTION

| *Data source* | *Nr* |
|---|---|
| Interviews: | |
| Healthcare personnel | 26 |
| Managers at different levels of HT1 | 6 |
| The EHR vendor | 3 |
| The app vendor | 4 |
| Regional implementation program | 1 |
| Regional ICT | 2 |
| Total | 22 |
| Observations: | |
| Observation at the pilot department at HT1 | 3 hours |
| Participation in meetings, workshops, and discussions. | 20 hours |

The data collection started a year after the initial implementation process and were conducted between June and October 2021. Most of the interviews with healthcare personnel were focus group interviews to avoid taking up too much of their time. With the other actors a mix of focus groups and single person interviews were done. The interviews lasted between 20 and 60 minutes. All interviews, except from one, were done on a distance, using Microsoft Teams for communisation. All interviews were recorded with a separate voice recorder and transcribed verbatim.

### B. *The NASSS-CAT Framework*

For evaluating the empirical implementation process, the NASSS-CAT framework was used as the basis for the interview guides and the analysis. Separate interview guides were designed for nurses, doctors, leaders, and IT/vendors to allow for domain-specific probing [9]. The NASSS-CAT framework is set to guide and evaluate the success or failure of technology deployments in organizations, as well as assessing complexity related to the implementation process [9][10]. Using the framework can help to predict and change directions during an implementation process to stimulate the success of an intervention [10][11]. The framework includes seven domains identified through systematic literature reviews and refined through empirical case studies of technology implementations [10][11][12].

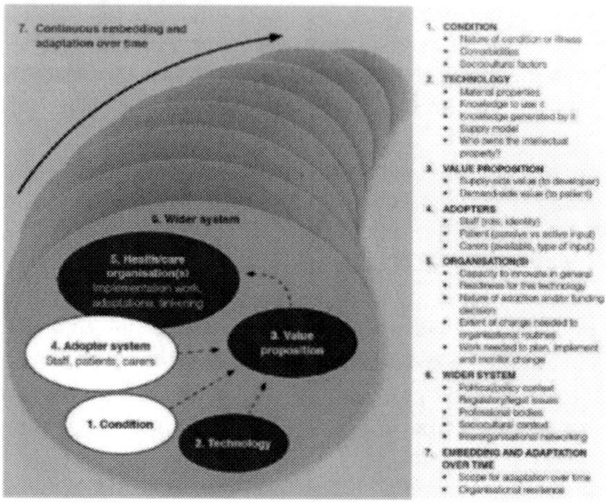


Figure 1. The NASSS-CAT framework

The seven domains are: I) the illness/condition, II) the technology, III) the value proposition, IV) the adopter system, V) the organizations, VI) the wider context, VII) embedding and adaption over time (see figure 1). The NASSS-CAT framework is set to guide and evaluate the success or failure of technology deployments in organizations, as well as assessing complexity related to the implementation process [9][10]. Using the framework can help to predict and change directions during an implementation process to stimulate the success of an intervention [10)[11]. The framework includes seven domains identified through systematic literature reviews and refined through empirical case studies of technology implementations [10][11][12].

By definition, the value proposition category addresses the actual value of an innovation and for whom it generates value. It questions whether a technology is worth developing in the first place and includes both the upstream values that follows the supply side logic of financial markets and investments decisions, including whether there is a straightforward and uncontested business case for generating revenue for the developer. It also addresses the downstream values that follow the demand-side logic of health technology appraisal, reimbursement, and whether there is strong and uncontested evidence that the technology is desirable for patients, effective, safe and cost-effective [12]. In accordance with the research question, which gives direction for the analysis of data we focus on the downstream values to answer our research question by addressing the values for the users and the organization. Further descriptions of the domains can be found in Greenhalgh et al. (2020) [12]. The value category therefore includes the consequences of the implementation, both intended and unintended ones.

### C. *Analysis*

The first step of the analysis was systematic reading and thematically coding of the interviews in line with the NASSS-CAT framework's seven domains. During this process, the content of the different domains was discussed and how they should be separate, since some of the findings related to more than one domain. For this paper, data representing consequences for the healthcare personnel and actors involved was of particular interest and thematically categorized in the "value proposition". However, consequences can also be understood as a secondary value or repercussions of the implementation and have a link to other domains as well. For this study, we divided the values proposition into 1) expected values (consequences) traced back to the HT1 list of outcomes, and 2) unexpected values (consequence) including the different repercussions. Both the expected and unexpected values were divided into categories represented by the different informants.

The next step was to produce narrative summaries for both the expected and unexpected values for all the groups of informants. During the analysis, there were three meetings with representatives from HT1 discussing both the definitions of NASS-CAT domains, the division of the value proposition, and the narrative summaries. The final step of the analysis was to merge the narrative summaries into overall themes to give insight into expected and unintended socio-technical values of implementing an app to support healthcare personnel's' point-of-care registrations of vital signs.

TABLE II. OVERVIEW OF OVERALL THEMES

| |
|---|
| A.The expected values of implementing the nursing app |
| B. Digitalising more paper-based processes |
| C. Improving the workpractice for doctors |
| D. Providing better overview for leaders, increased use of patient whiteboards |
| E. Improved communication between professions and organizations |

The themes are presented in Table II and elaborated on in the results part.

## IV. RESULTS

The focus of the paper is to outline intended and unintended consequences for the nurses by evaluating the implementation of an app for registering vital signs and NEWS score at the point-of-care. The aim is to generate an understanding of the repercussions the implementation generated both for healthcare personnel and the organization. In the results part the overall themes that emerged from the analysis are presented.

### A. *Intended Consequences of Implementing the Nursing App*

When deciding to implement the nursing app (illustrated in Figure 2), the management and the IT department defined some overall goals for the implementation.

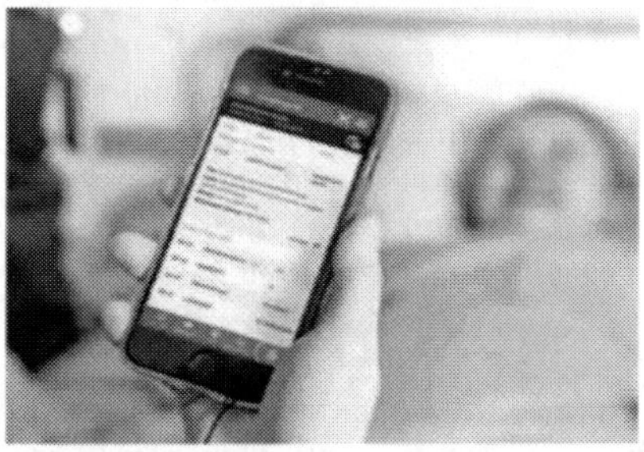

Figure 2. Illustration of the nursing app

These included improving the nurses' workflow and patient safety real-time registration, the possibility to share and reuse information electronically between different systems simultaneously, decision support for nurses, reducing paper-based processes and scanning of forms into the EHR. In addition, outlining and improving the digital competence for healthcare personnel by using mobile phones for clinical registration was important. There is a globally shared concern related to health personnel's digital competence, and skills are required in adopting new digital services [8].

When deploying the app, the IT department introduced it and demonstrated the use. The nurses instantly found it user-friendly, intuitive to learn and easy to use, due to the similarity between the app's interface and the existing paper form. The nurses particularly liked that the app automatically calculated the NEWS score at the point-of-care after filling in the vital signs. In addition, the app provided point-of-care decision support by suggesting which actions to take in accordance with the patient's updated NEWS score. Compared to before the implementation, the app provided the nurses an overview of all the patients' NEWS scores in the ward. Now, the nurses could check in the app, at the-point-of-care, if someone else had measured the vital signs recently, and check the patients previous score, instead of having to run around coordinating the work with each other. This vastly improved organizing v the workflow.

Due to the integration between the app and the existing IT infrastructure, the vital signs and NEWS Score were automatically transferred to the EHR system and the electronic patient whiteboards (Electronic whiteboards used in ward offices for coordination and overview of patients' conditions) at the wards, eliminating the need for double registration

Overall, the nurses found the app very useful, one of them stated: *"The Medanets app has only generated benefits for us. If a patient becomes ill, we can check the previous status on the phone, without having to search for the paper form. In addition, we don't have to double registrate or calculate NEWS anymore."*

Even if the app implementation fulfilled the intended consequences, some challenges were detected. First, for some nurses it was challenging to start using the mobile phone as a tool for registering vital signs. They needed extensive training in using the mobile phone as a clinical tool, before including the app in their daily practices. Some were concerned to make errors, or not be able to log onto the app. Second, some nurses reported that they experienced a barrier in using the mobile phone in front of patients. They described it as uncomfortable to use the mobile phone at the patient's bedside and worried that the patient found it impersonal and that it would hamper the communication with patients. They also worried that the focus on the digital work tool could risk lowering the focus on assessing the patient as a unique person. One of them said: *"There is a risk that especially young unexperienced nurses focus more on the digital results than the overall patient status."* Consequently, some nurses still noted the vital signs on paper and registered them in the app after leaving the patient room.

Another challenge was the mobile network at HT1. For instance, the Wi-Fi signal at parts of the hospital was very poor, especially at some of the wards. This resulted in instability in exporting data to the EHR and the electronic patient whiteboards. It was important to test and improve the mobile network in line with clinicians using the phone as a work tool Still, it was possible to register data offline in the app and exporting it whenever the WI-FI signal was back on. Sometimes, it was trouble with the mobile phones which made it impossible to log into the app. Occasionally, this led to going back to use the paper forms.

None of the nurses reported that using the app made them save time on measuring and filling in the vital signs at the point-of-care. However, in addition to the intended consequences summarised which the management and the IT department had defined as goals for the implementation the app generated several unintended consequences as well these are outlined in the following sub-sections.

### B. *Digitalising More Paper-based Processes*

From using the app for digital registration of vital signs, the nurses increasingly started to see the potential of digitalizing other paper-based processes as well to improve their workflow. In most wards, nurses reported that it was easy to propose new parameters or forms to integrate with the app. Healthcare personnel suggested improvements to their department leaders, which brought them further to the IT department. The IT department discussed the ideas and brought the doable ones to the app vendor.

The first new forms suggested by the nurses were the risk assessment forms related to the national "patient safety campaign" calculating risk for falling, bedsore, and nutrition screening. In addition, other parameters like weight, hight, and pain were included in the app. There was also an ongoing work on designing digital admission and submission forms for the app. Every time a new parameter or form was digitalized and included in the app, it increased the awareness amongst the nurse resulting in more systematic follow-up.

In the somatic department measuring vital parameters and calculating the NEWS score were well established routines. However, in some of the psychiatric and substance abuse wards they had just started using the NEWS procedure. Using the app became a means to increase awareness and more systematic registrations of NEWS. The implementation of the app also brought forth suggestions of other paper forms they found appropriate to digitalize, e.g., the violence measurement, symptoms and detoxication checklists. However, some of our informants outlined the importance of not digitalizing all forms. One of the doctors said: *"I don't feel the need to use such app for the scores we use in the IR, it would just generate a lot of clicking and extra time use for us."* This illustrates that not all work processes will be improved by digitalization.

During the first year of using the app, the IT department outlined that 100-150 changes were made to the app. They emphasized the importance of the app being flexible and adjustable in line with the general development of nurses' work practices.

### C. *Improving the Workpractice for Doctors*

The nursing app was not used directly by the doctors. However, it had an extensive impact on their workflow since the vital signs and NEWS scores were for the first time automatically exported from the app to the EHR system. The doctors could log in to the EHR and view results for patients assigned to them, which was of great importance when it came to following-up the patients with unstable medical conditions. The instant update in the EHR were also included in an overview of previous scores at different time intervals. Before the app was implemented the doctors had to walk to the ward and search for the medical chart in the ward office to get an overview of all the patients' NEWS scores. If they wanted to look at trends over time, they had to look through a pile of forms and compare the values manually. When they were busy and did not have time to search for the chart, they sometimes assessed the patient based on the approximate value based on their memory. One doctor stated that *"it improves the patient safety to be able to always have the updated NEWS value at hand."*

Before the morning rounds, the doctors had a meeting to assess all their patients and discuss further treatment. The quality of these meetings was improved from having updated NEWS scores at hand. Also following-up patients with unexpected NEWS values were much easier. Now they could look at the patients NEWS overview to assess if the current score was normal for this particular patient or if it was necessary to take further action. This was an asset to prioritising the right patients to focus on, resulting in more efficient patient treatment.

In addition, when the doctors were on call, they did "medical supervision" on patients at other wards. Now they were able to check the NEWS scores on these patients beforehand, enabling a more accurate assessment of patients they did not have the overall responsibility for and did not know so well. One doctor said: *"This is 2021 and it is natural that the information can be registered and read digitally, the world is moving forward, and we are all happy about that."* The managers at HT1 outlined that they did not expect the implementation of the app to have such extensive impact on the doctors' workflow.

*D. Providing Better Overview for Leaders – Increased use of the Patient Whiteboards.*

Besides being automatically exported to the EHR, the NEWS scores from the app were exported to the electronic patient whiteboards that all departments at the hospitals had in their ward offices (illustrated in Figure 3).

Figure 3.The patient whiteboard

The patient whiteboards presented the NEWS score as well as the values from the parameters in the "patient safety campaign" for each patient. One of the leaders said that they had not used the patient whiteboard actively for a while due to limited usefulness. However, after NEWS and vital parameters were included and systematically updated, they used it much more. The NEWS score for each patient came up with a colure code related to severity, and a countdown for when the patient needed the next score done.

Now the department leader could use the patient whiteboard to get an overview of the status of all their patients to plan the workday for the staff. If one nursing group had many patients with high NEWS scores demanding extra follow-up. Then, the leader could re-arrange the personnel in the different groups and transfer extra personnel for the group with patients that needed closely monitoring for that day. The leaders could use the patient whiteboard also to find out if some of the NEWS scores or other scores were not done or delayed and remind the nurses to do them. One department leader said: *"To have the NEWS score at the patient whiteboards improves our workday. We gather around the patient whiteboard every morning to plan the day and follow-up on scores that have not been conducted in time."* Some of the leaders expressed that it was challenging to get the nurses to use the patient whiteboards systematically since they already used the nursing app and the EHR system. However, these are important overview tools since everybody can access the same information whether it is the medical director, IT department, nurses, or doctors. The psychiatric and substance related department leaders stated that they did not use the patient whiteboards more after implementing the nursing app, since they did not have that many NEWS scores to conduct. The other parameters at the whiteboard were too oriented on somatic care to be of significant value for them.

*E. Improved Communication Between Professions and Organizations*

Improved communication was another important unintended consequence from using the nursing app. Since all vital parameters were registered at the patient's bedside, and the NEWS score automatically calculated, it was much easier to use the actual score in communications at different healthcare levels.

The communication between nurses and doctors improved from using the app because the nurses could use an updated NEWS score as an argument for following-up a patient, instead of using unspecific terms like "the patient looked a bit shabby, sickly, limp, pale." They felt more confident using the NEWS score when the app had calculated it automatically at the patient's bedside, than when they had to do it themselves. It was both more efficient and more accurate. One of the doctors said: *"it is much better to use objective parameters like NEWS score as the starting point for discussing the patient."* However, the doctors and nurses underlined that it is most important to look at the patient as a whole and not be blinded by the NEWS score. For some patients NEWS 8 might be normal and for some NEWS 0 could be dangerous. This depends on the total clinical picture and the dialogue between doctors and nurses for each patient.

Still, the main rule was to call the doctor when NEWS was above 5 to discuss the overall risk for the patient. One doctor said: *"Now we have the NEWS score instantly and can communicate using the same language."* This was assuring especially for unexperienced nurses and health workers that before may have observed that something was wrong with the patient without being able to confirm it, but now the changing NEWS score made it easier to explain the shift in the patient status. After NEWS was adopted, at first the doctors received a lot of extra calls related to the patient status, however now a year later the nurses used the app more as a supplement to the overall clinical assessment

In one of the psychiatric wards, they stated that using the app resulted in more systematic registration of NEWS, which in turn led to improved communication with the somatic department, in terms of increased focus on somatic deterioration in their patients. One nurse from the psychiatric department said: *"we have a common language now, when we talk about a NEWS score, the somatic doctors understands that this is something they have to follow-up on."*

Another communication improved was between the central hospital and the smaller district hospitals. One department leader at the district hospital said that earlier they had to argue a lot to get the patients transferred to the central hospital when their condition worsened. Now they could use a NEWS score value, e.g. of 9 or 10 and an overview of the patient's trends as an description of clinical deterioration and an argument for transferring a patient to the central hospital.

Consequently, the nurses found that using the NEWS score procedure systematically contributed to improve the quality of a patient's treatment and care in the different steps of the healthcare chain.

## V. Discussion

In the result section we presented the intended and unintended consequences that emerged from the implementation of the nursing app for registering NEWS in a hospital encompassing both somatic and mental health departments (see Table III).

TABLE III. AN OVERVIEW OF EXPECTED AND UNINTENDED VALUES FROM IMPLEMENTING THE NURSING APP.

| Overview of consequences | | |
|---|---|---|
| Intended consequences | Unintended consequences | Challenges |
| Intuitive and easy to use | Better overview for leaders through the patient whiteboard | Difficult to use for nurses with low digital competence |
| No doble registration | Better communication at different healthcare levels | Impersonal to use the phone at patient bedside |
| Improved workflow related to overview of NEWS | Increased digital focus – suggestions for more digital forms | Some still registered on paper first |
| Automatic calculation of score | Improved workflow for doctors | Less focus on overall assessment |
| Instant export of NEWS to EHR and patient whiteboard | | Problems with the phones and the network |
| Less paper-based work | | Not very time saving |

The intended consequences were in line with the overall aim of implementing the app in clinical practices. The study demonstrates that the implementation of the app was interpreted as a success by the nurses despite that some worry that using technology at the patient bedside shifts the focus from the patient to, in this case, the phone. Healthcare personnel experienced it as useful and intuitive. The app improved their workflow and to the end, it improved the treatment and care of patients also. The app fulfilled the expectations of fast and efficient communication of vital signs that increase accurate and timely documentation of vital signs, which in turn leading to safe patient treatment and care. In addition, the app gave the nurses clinical decision support at the point-of-care that improved their clinical confidence and improved the communication about the patient's situations with the doctors [1][2][5].

Defining an implementation of a digital solution for healthcare a success is in itself a very rare and important finding. Often such implementation demands for comprehensive training before using the solution, and resource-intensive adaptation of both technology and workflow to make the technology work optimally. In addition, technology often demands for workarounds and fulfils only parts of healthcare organizations' requirements. To solve several requirements for a heterogeneous group of different healthcare personnel, the system implemented is often a large, complex digital solution, like an EHR system, which takes time to adapt to clinical practice.

However, this app is the opposite, it is a simple app designed to solve a clearly defined work task for one specific group of healthcare personnel. The digital form in the app is identical to the paper-based form, hence the workflow of measuring the vital signs is the same, only the registration process has changed. As a result, the nurses stated that they did not find it time saving to use the app for registration of vital signs. This is in line with previous studies, for example Dall'Ora et al. [2] argues that there is no evidence that nurses save time when using electronic vital signs recording. Our study did not measure time related to using the app for recording and calculating NEWS score. However, this study identifies other values generated by the app. It increases the quality and security of the work processes by offering decision support based on NEWS values, automatically calculating the score, instantly exporting the results to the EHR, and erasing the double registration and the risk of misunderstanding bad handwriting. The app also improves the work processes surrounding the actual registration. This includes the collaboration amongst nurses, the need for less resources to scan the paper forms and the possibility of reusing NEWS to the EHR and the patient electronic whiteboards. The managers at HT1 were aware of the risk of not making measuring vital parameters more efficient, however, the surrounding values were extensive enough for the process to be defined as successful.

One of the challenges of implementing new technology reflects the diversity of digital competence among nurses [8]. The management at HT1 saw the implementation of the app as a possibility to get an overview of the digital competence amongst their workers. and how they experienced using handheld devices at the point-point-of care. Kaihlanen, et al. [8] outlines the overall concern related to nurses' digital competence since nurse's informatics competence affects the quality of healthcare [8].

It is important to raise the digital competence of the nurses without compromising the clinical assessment. More and more of the digital solutions implemented in healthcare includes digital solutions for registering data. The EHR vendor for instance has worked for years to develop mobile solutions for accessing the EHR and registering data. Other vendors also want to transfer functionality to mobile solutions, and there is a wave of apps entering the healthcare. However, using all kinds of digital solutions, replacing the overall assessment is impossible, the experienced nurses said that this was important to remember in a busy workday. To enable more digital solutions for healthcare personnel it was important for the management to to get an overview of the capacity for the mobile network at the health trust. Within the group of

Still, increasing the digital competence of the users and expanding the use of digital solutions demands for an up-to-date mobile infrastructure to make sure that the phones always are available at all locations. If using the phone is too cumbersome, in terms of e.g., no network, out of battery, or delays related to logging in, there is a risk that the nurses will stop using the mobile solution and go back to their old paper-based routines. In this case, the importance of the nursing app to be more efficient and easier to use than paper was outlined as a precondition to use it in the first place. However, the regional IT department ran the IT network, and scaling up the mobile use to clinical use for an entire health trust was new for them and required some improvement and optimizing of their mobile infrastructure.

The implementation of the app implied unintended consequences. These are of particular interest to discuss for understanding the socio technical consequences the implementation of the app brought about. First, the app generated new practises and workflows both for leaders and doctors, who did not fuse the app, but who were influences by receiving data from the app in other systems. Both groups report that improved overview and patient safety were important values generated by the app. In relation to the increased use of the patient electronic whiteboards, the results underscore the importance of engaged and digitally mature leaders. In departments with such leaders, the patient electronic whiteboards went from being a rather passive and random used work tool, to an important means for coordinating work practice and resource allocation at the ward.

Existing research points out a lack in standardized clinical processes for measuring vital signs, and inadequate description of the vital signs' observation methods used [2]. HT1 had standardized the process of measuring vital signs by using the NEWS Score years before implementing the app. However, this study also showed how a systematically and standardized procedure for measuring and registration of vital signs and NEWS Score improved the communication between different healthcare professions, healthcare specialities, and the hospital's clinical departments. A standardized language for communication (NEWS), generated more confidence amongst the nurses when reporting the results to the doctors. In addition, the doctors could evaluate the patient not just by the status today, but also the overview from several days resulting in improved patient safety.

## VI. Conclusion and Future Work

Implementing the nursing app at HT 1 was an overall success both for healthcare personnel and the organization, due to the instant usefulness for nurses as well as the repercussions including positive consequences for doctors, leaders and the improved communication across healthcare.

To define healthcare technology implementations as a success, analysis framework like NASSS-CAT is very useful for to make it clearer what are the intended and the unintended consequences, and how are they fulfilled in the process. Using the framework, revealed that the app generated several intended and unintended consequences that were important to exploit further to understand the repercussions this simple app implementation generated.

Lessons learned from the study:

- The app did not generate extensive time saving for nurses, however other important consequences made the implementation a success.
- Implementing a simple app can generate several unintended values with just as much benefit for the users and the organization as the intended ones. It is important that the organization can rapidly adjust to, and embrace, the unintended values.
- Implementing such simple app and increasing the digital competence and digital interaction between the systems in their IT ecosystem, made the

organization and the healthcare personnel better prepared for forthcoming more complex digital solution and the possibility for exploiting the digital possibilities for optimizing healthcare increased extensively.

When implementing digital solutions in healthcare it can be difficult to measure cause/effect values related to time saving and reduced costs during the implementation process. However, conducting evaluation research generates an increased long-term learning effect for organizations as well as for knowledge production.

Future work will include the relation between the app and the EHR system in related to how large EHR system can keep up with the fast-evolving apps and the interrelation between those different technologies in the IT infrastructure. This will demand for an extensive understanding of the IT infrastructure and the organization at HT1. It is also important to not include to much functionality in the app, to avoid for it to become too complex.

## REFERENCES

[1] M. Yeung, S. E. Lapinsky, J. T. Granton, D. M Doran, J. A Cafazzo. "Examining nursing vital signs documentation workflow: barriers and opportunities in general internal medicine units." Journal of clinical nursing, vol. 21, no. 7-8, pp. 975-982, 2012.

[2] C. Dall'Ora, et al., "How long do nursing staff take to measure and record patients' vital signs observations in hospital? A time-and-motion study." International journal of nursing studies, vol. 118, pp. 103921, 2021.

[3] M. Churpek, R. Adhikari, and D. Edelson, "The value of vital sign trends for detecting clinical deterioration on the wards." Resuscitation, vol.102: p. 1-5, 2016.

[4] Royal college of Physicians, National Early Warning Score (NEWS) 2. 2022.

[5] O. C. Redfern, P. Griffiths, A. Maruotti, A. R. Saucedo, and G. B Smith, "The association between nurse staffing levels and the timeliness of vital signs monitoring: a retrospective observational study in the UK." BMJ open, vol. **9**, no. 9, e. 032157, 2019.

[6] J .E. Ball, T. Murrells, A. M. Rafferty, E. Morrow, and P. Griffiths, "Care left undone'during nursing shifts: associations with workload and perceived quality of care." BMJ quality & safety, vol. 23, no. 2, pp. 116-125, 2014.

[7] Nordlandssykehuset. Mer om Nordlandssykehuset. In english: More about Nordlandssykehuset 2020, [retrieved: January 2022], Available from: https://nordlandssykehuset.no/om-oss/om-oss-#mer-om-nordlandssykehuset.

[8] A. M. Kaihlanen et al., "Nursing informatics competences of Finnish registered nurses after national educational initiatives: A cross-sectional study." Nurse Education Today, vol. 106, pp. 105060, 2021.

[9] T. Greenhalgh, et al., "Beyond adoption: a new framework for theorizing and evaluating nonadoption, abandonment, and challenges to the scale-up, spread, and sustainability of health and care technologies." Journal of medical Internet research, vol. 19, no. 11, pp. e8775, 2017.

[10] T. Greenhalgh, H. Maylor, S. Shaw, J. Wherton, C. Papoutsi, V. Betton, N. Nelissen, A. Gremyr, A. Rushforth, M. Koshkouei, and J. Taylor, "The NASSS-CAT tools for understanding, guiding, monitoring, and researching technology implementation projects in health and social care: protocol for an evaluation study in real-world settings." JMIR research protocols, vol. 9, no. 5, pp. e16861, 2020

[11].S. Abimbola, B. Patel, D. Peiris, A. Patel, M. Harris, T. Usherwood, and T. Greenhalgh, "The NASSS framework for ex post theorisation of technology-supported change in healthcare: worked example of the TORPEDO programme." BMC medicine, vol. 17, no. 1, pp. 1-17, 2019.

[12] T. Greenhalgh, and C. Papoutsi, "Studying complexity in health services research: desperately seeking an overdue paradigm shift.." Springer, pp. 1-6, 2018.

# Use of Electronic Tools in Norwegian FACT Youth Teams: A User Perspective

Erlend Bønes, Conceição Granja, Terje Solvoll
Norwegian Centre for E-health Research
University Hospital of North Norway
Tromsø, Norway
Faculty of Nursing and Health Sciences
Nord University
Bodø, Norway
email: {erlend.bones, conceicao.granja, terje.solvoll}@ehealthresearch.no

*Abstract*— **Flexible Assertive Community Treatment (FACT) youth teams deliver integrated services to youths aged 12-25 with mental health issues. There has been reported challenges with use of e-health solutions for FACT teams targeting an adult population. The objective of this study was to examine challenges with e-health solutions for FACT youth teams, with a special focus on the electronic whiteboards they use. We did semi-structured interviews in 3 Norwegian FACT youth teams. We identified challenges with the electronic whiteboards, electronic health records and team calendar solutions. There is a need for improved e-health solutions for FACT youth teams in Norway.**

***Keywords- FACT; FACT youth; mental health; electronic whiteboard; electronic health records; e-health.***

## I. Introduction

Flexible Assertive Community Treatment (FACT) is a model for delivering integrated services to persons with long-term mental illness [1]. The model was developed in the Netherlands in the early 2000s and is a variant of Assertive Community Treatment (ACT). While ACT is designed to provide continuously intensive health services to all their patients, FACT is targeted at patients that require intensive services in some periods, but less intensive in others [2]. FACT teams work with a shared caseload on patients when they require intensive services, and individual case management when the patients are stable [1]. This makes FACT better suited to areas with lower patient populations, that do not have a high number of patients that requires continuously intensive health services.

Most FACT teams target adult patients, but some teams target youths. The Netherlands have had FACT youth teams since 2005, and there are around 60 FACT youth in the Netherlands today [3].

Traditional health services in Norway have had problems reaching youths with complex issues. Cooperation between levels of health care is affected by unclear responsibilities, and there is a lack of integration of services [3]. The goal of FACT youth teams is to meet these challenges, by providing integrated services for their patient population [3]. The target group for Norwegian FACT youth teams are youths aged 12-25 years old. FACT youth teams are multidisciplinary and should consist of a team leader, team coordinator, child and youth psychologist, child and youth psychiatrist, family therapist, user specialist, and a work/education specialist. In 2021, there were 3 FACT youth teams in Norway, and around 70 FACT teams targeting adults.

In Norway, specialist mental health services are the responsibility of the government. The services are delivered by hospitals or community mental health centers, that are owned by one of the four Regional Health Authorities. Primary care and local services are delivered by the 356 municipalities. Most FACT teams in Norway are organized as a cooperation between specialist care and one or more municipalities [4].

Standardized patient pathways were introduced for all mental health services in Norway in 2019 [5]. The goals of the standardized patient pathways were to reduce variance in treatment, ensure user participation, and improve coordination between various health services. Thus, all patients in Norwegian FACT teams should be in a standardized patient pathway.

FACT teams have daily meetings, where they discuss status and plan further follow-up of patients that require intensive treatment [1]. The teams use an electronic whiteboard to display an overview of these patients. This makes the electronic whiteboard one of the most important tools for the FACT teams. Patients who receive case management are discussed less frequently by the team, usually once a week.

Electronic health records (EHRs) are an important tool for health care workers to get relevant information about their patients and document treatment the patient get. Specialist health care in three of the four Norwegian health care regions use the EHR system DIPS AS, while the fourth are using DocuLive provided by Siemens AS. There are several different EHR systems in primary care.

There have been reported challenges with information and communication (ICT) solutions for FACT teams targeting adults in Norway. This includes issues with electronic whiteboards, EHRs and calendars [4][6]. The goal of this study was to examine FACT youth teams perspectives on electronic whiteboards and other ICT solutions in

Norwegian FACT youth teams. These results can be used to inform how an electronic whiteboard solution for FACT youth teams should be designed.

The rest of this paper is organized as follows. Section II describes the methods used. Section III shows the results. Section IV presents the discussion. Section V describes the conclusions, including future work.

## II. METHODS

Three Norwegian FACT youth teams were included in the study. The teams chosen were the 3 FACT youth teams operating in Norway at the time of the study. Team 2 and 3 was organized as a cooperation between primary care and specialist care. Team 1 was established by a municipality, and was therefore mainly based in primary care, but they had a psychologist employed by specialist care. All 3 teams were situated in urban areas. Table 1 shows the characteristics of the teams included.

Building on a Computer supported cooperative work framework [7][8], we conducted semi-structured interviews with the teams. The interviews were conducted during the fall of 2021 by one researcher. We invited all members of the teams to participate in the interviews, but for various practical reasons not all team members were present during the interviews. The number of participants in each interview was from 1 to 3, for a total of 7 participants. We used an interview guide and presented predefined use cases as a starting point for open-ended discussions about use of the electronic whiteboard and other ICT solutions.

The interview guide focused on how an ideal electronic whiteboard would look, if the electronic whiteboards should be integrated with the EHR or other systems, and if it would be useful to be able to extract statistics from the electronic whiteboard. See Fig. 1 for the list of interview questions. The use cases showed different scenarios where the electronic whiteboard or other ICT solutions were involved. The use cases discussed during the interviews were referral of patients to FACT youth teams, use of the whiteboard during the daily meetings, updating of the whiteboard after meeting a patient, transfer of patient from intensive follow-up to case management, end of treatment of patient from FACT youth team and use of the team calendars. See Fig.2 for the use cases discussed. The interview guide and use cases were developed based on experiences from a previous study of ICT solutions for FACT teams for adult patients [9].

TABLE I. CHARACTERISTICS OF THE TEAMS

| Characteristics of the teams | | | |
|---|---|---|---|
| ***Team*** | ***Coverage area*** | ***Team organization*** | ***Number of interviewees*** |
| 1 | One municipality | Mainly primary care | 3 |
| 2 | One city district | Primary and specialist care | 3 |
| 3 | One city district | Primary and specialist care | 1 |

- How do you use the electronic whiteboard today?
- Do you want to use the electronic whiteboard for other purposes?
- Is there any kind of integration between the electronic whiteboard and EHR solutions? If no, is this something you want?
- What team members should have access to the electronic whiteboard?
- Do you have a need to extract reports and statistics from the electronic whiteboard? If yes, what kind of reports and statistics?
- What calendar solutions do the teams use? Is there a need for better calendar solutions?

Figure 1. Interview guide questions

The interviews were recorded and transcribed. In addition to this, the researcher took some notes during the interviews, and wrote a memo after each interview. This memo contained ideas and thoughts from the interview. One of the researchers read the transcripts and memos several times. Based on this we developed preliminary themes of data for the data analysis.

## III. RESULTS

The themes of data identified in the preliminary analysis of the data matched the main ICT solutions used by the teams; electronic whiteboards, calendars and EHR.

The electronic whiteboards used by the teams were standalone solutions made in Microsoft Excel. The electronic whiteboards did not have any integration to the EHR or other systems. All 3 teams reported that they would like the electronic whiteboard either integrated to the EHR or as a part of the EHR. Team 2 said that because of the lack of integration, the electronic whiteboard is mainly a tool for coordination, and they still must do all the documentation in the EHR. Team 3 said that it was important to have the same diagnoses in the EHR and electronic whiteboard, since the teams often work with tentative diagnoses.

Team 1 reported that they wanted the whiteboard to be more oriented towards the users' family and network, since this is an important part of how FACT youth teams work.

The interviewer asked the teams if it would be useful for them to extract statistics about their users from the whiteboard solution. Team 1 said that they had no large need of this, because they had a good overview of their patients due to a small number of patients. Team 2 wanted to be able to extract statistics. The reason for this was that they wanted to justify what they did and the results they were getting. Ideally, they wanted to use the statistics to estimate how many inpatient days they prevented for their patients. Team 3 also said that they wanted to be able to extract statistics. They wanted information about the number of patients, number of patients on compulsory treatment, number of

- Referral of patients to FACT youth teams
- Use of the whiteboard during the daily meetings
- Updating of the whiteboard after meeting a patient
- Transfer of patient from intensive follow-up to case management
- End of treatment of patient from FACT youth team
- Use of the team calendars

Figure 2. Use cases discussed

patients with specific diagnoses and number of patients who get follow-up from child welfare services.

The teams also pointed out that they wanted the electronic whiteboard to highlight deadlines related to standardized patient pathways and treatment plans. Team 2 wanted deadlines related to treatment plans updated automatically in the electronic whiteboard. Team 3 said that they are in favor of the implementation of standardized patient pathways in Norway, but it can be a lot of work to keep track of where the patients are in the pathways.

The teams also reported some issues with the use of calendars. One issue is that they have several different calendars they use, this includes calendar in EHR for specialist care and Outlook calendars for specialist and primary care. The calendars available to each team member was based on their employment. Team 1 said that they absolutely have a need for a common calendar, that would make it easier to coordinate a shared caseload. They would also like to have the calendar functionality connected to the whiteboard. Team 2 reported that the calendar in the EHR for specialist care is hard to use on mobile devices. Team 3 said there is a simple overview of plans for the current week for patient on their whiteboard. This overview be seen as a calendar for the patient. Some adult FACT teams have reported that calendars are useful for safety purposes [9]. The FACT youth teams reported that this was not relevant for them at this time. Team 1 said that this is not relevant for their current patients. Team 3 said that their patients receive voluntary treatment, and that the team do not intrude on the patients, so this is not an issue for them.

One dilemma that several teams brought was how information regarding family members should be written in the EHR. Health issues of the other family members might be related to the youths' issues. However, since it is only the youth that is the teams' patients, health information about other people should not be written in their EHRs.

## IV. Discussion

An important issue reported by all three teams is a lack of integration between the electronic whiteboards and the EHR systems. An ideal solution would allow information from the EHR systems to be directly included in the electronic whiteboard. This could include patient IDs, diagnoses, and other relevant information. Data from the whiteboard should also be transferred to the EHR. In a study of FACT teams for adults [9], it was shown that also adult teams wish integration between the electronic whiteboards and EHR.

One team wished that the electronic whiteboard should be more focused on the patients' family and network, to reflect the way the teams work. Some whiteboard solutions already contain basic text fields for information about family and network. However, this solution might be improved upon. How the electronic whiteboards can be better adapted to working with the patients' family and network could be a topic for further research.

Two of the teams stated that they wanted to be able to extract statistical information from the whiteboards about number of patients and diagnoses. The statistics could be used for various administrative purposes, including being better able to justify the work the teams do for their funders. A solution that shows basic statistics about patients and diagnoses should be straightforward to implement in an electronic whiteboard solution. One team also wished to be able estimate how many inpatient days they save. Investigating if a whiteboard solution can assist in this beyond providing basic information could be a topic for further research.

One issue related to use of the EHR is the dilemma about what should be written in the EHR about family members of the patient. The health issues of family members might be relevant to the issues of the youths. However, it is only the youth that is a patient for the FACT youth teams. It is unlikely that this challenge can be solved directly by an ICT solution, but it is an issue with the use of the EHR we think it is important to highlight.

The teams reported that they use several different calendars, but at the same time also have some issues finding information about the availability of other team members. These results are similar to results from FACT teams targeting adults [9]. A common calendar for the whole team could be a useful solution. However, this requires that the new calendar becomes the preferred solution and not just another calendar that adds to the confusion. Since the electronic whiteboard is the daily tool of the teams, it would be natural to connect the calendar to the whiteboard.

The teams also said that they wished that the electronic whiteboard displayed deadlines for standardized patient pathways and treatment plans. This would make it easier to for the teams to keep track of the different deadlines. This shows that the teams have many different deadlines to administer, and they wish that the whiteboard help them keep track of these.

Many of the issues we found about FACT youth teams are the same as issues found in FACT teams for adults [9]. This includes challenges with the electronic whiteboard and calendars. One apparent difference between teams for youths and adults is a higher emphasis on family and network for youth teams. Electronic whiteboards for FACT teams for adults can also display information about family and network, and involving family members is often important for these teams. However, the emphasis on this is higher in FACT youth teams, and FACT youth teams have some unique issues, like the dilemma regarding family members and the EHR as described above. Despite these differences, we believe that an improved electronic whiteboard could be designed to serve both FACT youth teams and FACT teams for adults.

### A. *Study limitations*

There were only 3 FACT youth teams included in this study. These were all 3 FACT youth teams that was operating in Norway at the time of the study. Because of this, these results should be seen in conjunction with results for FACT teams targeting adults, while being aware of any differences between the types of teams.

## V. CONCLUSION

This study shows that FACT youth teams have several issues with their current electronic whiteboards and other ICT solutions. Many of these issues are the same as issues found in FACT teams targeting adults [9]. Better ICT solutions are needed for both types of teams.

In the future we will analyze these results together with results regarding electronic tools for FACT teams targeting adults. Together, they will form the basis of a detailed description of requirements for e-health solutions for FACT teams. Any differences for requirements for teams targeting youths and adults will be highlighted.

## ACKNOWLEDGMENT

We would like to thank the FACT youth teams participating in this study. We would also like to thank Trond Hatling at the National Resource Center for Community Mental Health (NAPHA), for helping us get in touch with the teams.

The study has been funded by the Norwegian Directorate of Health and the Research council of Norway.

## REFERENCES

[1] J. R. van Veldhuizen, "FACT: a Dutch version of ACT," Community Ment Health J, vol. 43, pp. 421-433, May 2007.

[2] G. R. Bond and R. E. Drake, "The critical ingredients of assertive community treatment," World Psychiatry, vol. 14, pp. 240-242, June 2015.

[3] Norwegian National Advisory Unit on Concurrent Substance Abuse and Mental Health Disorders, "FACT Ung Modellbeskrivelse," [FACT youth model description], February 2022. (In Norwegian).

[4] A. Landheim and S. Odden, "Evaluering av FACT-team i Norge – Sluttrapport " [Evaluation of FACT teams in Norway], April 2020. (In Norwegian).

[5] M. Ådnanes, J. R. Høiseth, M. Magnussen, K. Thaulow, and S. L. Kaspersen, "Pakkeforløp for psykisk helse og rus–brukere, pårørende og fagfolks erfaringer," [Standardized patient pathways for mental health and addiction - the experiences of users, relatives and professionals], January 2021. (In Norwegian).

[6] E. Bønes, C. Granja, and T. Solvoll, "Implementation of the Flexible Assertive Community Treatment (FACT) Model in Norway: eHealth Assessment Study," Journal of medical Internet research, vol. 24, pp. e32220, January 2022.

[7] G. Fitzpatrick and G. Ellingsen, "A review of 25 years of CSCW research in healthcare: contributions, challenges and future agendas," Computer Supported Cooperative Work (CSCW), vol. 22, pp. 609-665, June 2012.

[8] W. Pratt, M. C. Reddy, D. W. McDonald, P. Tarczy-Hornoch, and J. H. Gennari, "Incorporating ideas from computer-supported cooperative work," Journal of Biomedical Informatics, vol. 37, pp. 128-137, April 2004.

[9] E. Bønes, C. Granja, and T. Solvoll, "Use of e-health in Norwegian FACT teams: a user perspective," Medical Informatics Europe (MIE 2022), in press.

# Technical Viewpoint of Challenges, Opportunities, and Future Directions of Policy Change and Information-Flow in Digital Healthcare Systems

Areeg Samir
*Computer Science Department*
*UiT The Arctic University of Norway*
Tromsø, Norway
email: areeg.s.elgazazz@uit.no

Håvard D. Johansen
*Computer Science Department*
*UiT The Arctic University of Norway*
Tromsø, Norway
email: havard.johansen@uit.no

*Abstract*—**Digital healthcare systems often run on heterogeneous devices in a distributed multi-cluster environment, and maintain their healthcare policies for managing data, securing information flow, and controlling interactions among systems components. As healthcare systems become more digitally distributed, lack of integration and safe interpretation between heterogeneous systems clusters become problematic and might lead to healthcare policy violations. Communication overhead and high computation consumption might impact the system at different levels and affect the flow of information among system clusters. This paper provides a technical viewpoint of the challenges, opportunities, and future work in digital healthcare systems, focusing on the mechanisms of monitoring, detecting, and recovering healthcare policy change/update and its imprint on information flow.**

***Keywords—Digital Health; Healthcare; Distributed Environment; Policy; Multi-Cluster; Monitor; Detection; Performance; Workload; Recovery.***

## I. Introduction

Digital healthcare systems deliver services to consumers and patients and help them manage their health by providing a real-time communication environment. Each system is commonly organized into many clusters with different capacities, configurations, resources, and policies. A cluster has a set of components that include several services. For instance, a healthcare component might include services, such as primary and hospital care services. Each service consists of workflows, information pathways, and processes. A service could be integrated within the same organization to create a single unit. Different systems are also used for various purposes; for example, the municipality uses one electronic medical record system for documentation, while the hospital and primary healthcare services use another type. Moreover, various resources are shared among multiple organizations and healthcare individuals. Here, it is difficult to predetermine a fixed set of system individuals in such a dynamic environment as their roles and access control could be changed with system policy change. The system is expected to scale to respond to load variations, leading to unexpected overhead due to data movement and high resource consumption impacting system services, processes, and information flow.

Different distributed management information system policies of healthcare are applied to manage the system's information flow, such as data security, data management, health information dissemination, healthcare system resources, and data analysis. In such a system, communications, dependency, precedence, information shared between the system's clusters, different data standards, and multiple processes might be impacted by changing/updating policies associated with data flow in the system.

The lack of integration between the system's clusters and services complicates sharing, accessing, and flowing of information in the system. Coordination is more of a challenge due to the increasing complexity of services and the increasing complexity of their political environment.

The paper aims to explore challenges, opportunities, and future research directions in the digital healthcare system focusing on the impact of policy change/updates on the flow of information and its propagation at different system levels. We provide high-level information about the paper topic while being low-level enough to represent several key research areas that mainly focus on the challenges and opportunities within digital healthcare systems. We mainly concentrated our investigation on the perspectives of sharing, monitoring, detecting, and recovering the policy change and information flow within healthcare system processes.

The paper is organized as follows. Section II explains the method of research. Section III explores the information flow of the digital healthcare system. Section IV discusses the challenges of information flow within a digital healthcare system. Section V introduces opportunities. Section VI provides a set of future research directions followed by conclusions section.

## II. Research Method

The paper's objective is addressed by answering a research question: what are the challenges and opportunities that the digital healthcare system might face when healthcare policies change in a distributed environment?

### A. Search and Review Strategy

We conducted a systematic review according to PRISMA guidelines [1] to identify documents that reported on the policy change and information flow in healthcare and digital healthcare systems. We reviewed the state-of-the-art focusing on the policy's change/update and its impact on the information flow in the digital healthcare domain. We are interested in investigating various mechanisms that monitor, detect, and recover the change and its impact on the flow.

Moreover, the research paper focused on reports issued from pioneer healthcare organizations World Health Organization (WHO), Joint Commission International, and the Organization for Economic Co-operation and Development (OECD) for the period from 1999 to 2022 for several reasons:

- Detailed quantitative data on the use of digital health appeared during that period.
- WHO released its first Guidelines [2] on digital health.
- A significant development in digital health became active during that time [3].
- Significant relevant research was conducted during that period [4]–[10].
- Diversity of digital health strategies [11].

According to the previously mentioned organizations, we selected reports and documents issued by the healthcare associations focusing on healthcare in general and digital healthcare in specific, and we classified them into regions according to the published classification by WHO as shown in Table I. After that, we selected around 19 countries from all regions[1], and we investigated the healthcare system use-cases of these countries. The selection is made according to the common features, the characteristics of their healthcare, and digital healthcare systems within and across regions in which we are interested, such as:

- Digital healthcare technology: the ways that health individuals communicate, access/store/process medical and health records, research health information, and engage in a person-to-person exchange of text, audio, video, and other data.
- Digital healthcare processes: the workflows and information pathways within and across the organization.
- Multiplicity of policies: methods followed in handling various policies with their versions in the organization.
- Resources variations: various resources in varying amounts and configurations are dispatched towards the system's activities depending on the dynamic requirements. The scale of resources is the basis of the diversity of the actions taken.
- Dynamic adaptation: adaptation to hazards and adequately environment conditions and to the emerging needs and requirements of the situation.
- Collaboration across and within an organization: the activities are performed by organizations from different sectors, including interactions, operations, and relationships of system components within and across the organization.
- Consistency of information sharing across the whole process of the system.

[1]Europe: Nordic countries, Netherlands, Czech Republic, Serbia. Asia: Japan, China, Vietnam, Taiwan, Kurdistan. Africa: South Africa, Tunisia, Egypt, Nigeria, Zambia, America: the USA

TABLE I
HEALTHCARE ASSOCIATIONS ACCORDING TO REGIONS

| Region | Associations |
|---|---|
| Europe | Public Health Association - European commission |
| | Council of Europe and Health |
| America | Pan American Health Organization (PAHO) |
| | American Public Health Association |
| | Joint Commission on Accreditation of Healthcare Organizations (JCAHO) |
| | National Institutes of Health (NIH) |
| Asia | The Association of Asian Pacific Community Health Organizations (AAPCHO) |
| Africa | Africa Health Organisation (AHO) |
| | West African Health Organisation (WAHO) |
| | Amref Health Africa |
| | African Union |

### B. Search Criteria

Moreover, papers are selected from major journals and conferences[2] in health, healthcare, digital healthcare, and medical care. Based on that, we searched Web of Science, PubMed, Scopus, Medline, Scielo, IEEE Xplore Library, ACM Digital, Science Direct, Springer Link, and Google Scholar using criteria: years = 1996 – 2022, and keywords = "Healthcare System Information Flow" OR "Information Flow Control Multi-Cluster System" OR "Policy Change and Update Management" OR "Monitor Policy Change" OR "Detect Policy Change" OR "Recover Healthcare System" OR "Health Care Management" OR "Healthcare Service Quality" OR "Information Share" OR "Hierarchical Policy Management" OR "IoT in Healthcare" OR "Cost Reduction" OR "Healthcare Resource Allocation" OR "Hospital Workflow Processes" OR "Uncertainty in Healthcare Management" OR "Digital Healthcare Security Cloud".

### C. Search Outcome and Analysis

We ended up with more than 2000 papers distributed unevenly among the journals and conferences.

We created a map analysis of the "Digital Healthcare" definition based on bibliographic analysis utilizing VOSviewer version 1.6.18 to concentrate on the works in digital healthcare during the period mentioned above. We determined whether each paper: (1) had some form of qualitative/quantitative analysis, technical viewpoints, method development, or review of methods in the keywords mentioned above, and (2) considered a multi-cluster approach. We conducted a systematic literature review of potential barriers to policy change and information flow in digital healthcare systems. That left us with 241 papers for a full review. We evaluated the papers according

[2]healthcare, health policy management, health services research and policy, transactions on software engineering, computing transactions, Internet of Things, International Journal of Trend in Research and Development, Journal of Business & Economic Policy, International Journal of Advanced Computer Science and Applications, International Conference on Pervasive Computing Technologies for Healthcare

to different criteria: information flow control modeling, dynamic monitoring, continuous detection, data sharing, policy consent, security concern, information flow recovery, and cost reduction. Papers that did not focus on those criteria were excluded. We ended up with 22 papers that are relevant for our survey.

### *D. Search Boundaries*

We analyzed the healthcare, digital health, and medical systems in cloud computing, edge computing, statistical and dynamic analyses, machine learning, telemedicine, e-Health, security (blockchain platforms), and IoT. Due to the enormous scope of digital health technologies and literature studies, the paper could not discuss all aspects of these fields. The paper aims to discuss the recent challenges and advances in healthcare and digital healthcare.

## III. Information Flow in Healthcare System

According to the research method conducted previously, the infrastructure of a healthcare system consists of various electronic medical/health records, databases, networking technologies, and other specific biomedical, administrative and financial technologies that generate, transmit and store healthcare information [4][6][7][12]–[16]. The information flow from healthcare providers (Health Individuals) is entered into an Electronic Health Record (EHR) [17] and then networked to regional and national databases through EHR [12][15][18], as shown in Figure 1.

Based on that, a healthcare provider might generate various complex processes that affect the flow of information within a system [8][12][15][19][20]. The number of possible processes increases with the number of interacted participants. Thus, the complexity of processes varies among the interacted healthcare participants. For example, four healthcare participants in a hospital care team might create eight separate processes or more, as shown in Figure 2. The main flow of information among the healthcare participants is within two units: inpatient care (patient and admission to medical department) and primary care (doctor, medical department, general practitioner). Both units share information, and only authorized participants such as medical staff could access the collected data. The patient needs a general practitioner referral (A) for specialist treatment coverage, communication with laboratories and radiology services, and sick leave. The patient uses the referral (B) to the general practitioner's acute-care services and makes co-payments directly to care providers during their visits. The general practitioner admitted the patient to the medical department (C) to be examined and treated by a specialized doctor (D:H). At the same time, the general practitioner transmits prescriptions electronically to pharmacies and uses electronic health records to store, access, and retrieve patient data. Such flow of information is shared between the system's participants, and it might constitute one or more processes that might include other sub-processes. For example, a nurse directs a patient to a medical doctor who takes information from the patient and records it, see Figure 2.

When referring the patient, a doctor sends a referral message to the Health Registry Server. The message is stored in a Repository (e.g., health, clinical, national, and statutory). The doctor sends a referral message to a requested hospital, which sends the patient's information to the Registry Server for future use. The flow of information could be direct flow, such as the interaction between patients and physicians (e.g., doctor and nurse), as shown in Figure 2. It could be an indirect flow through central units such as funds and insurance companies, as shown in Figure 1. In such a case, the hospital is responsible for refunding and defining its policy.

The flow could be more divergent such as a patient needs unavailable service in a hospital. In such a case, the hospital would buy the service from another specialized hospital, which generates a net of interacting hospitals with various processes and sub-processes, including their dependency and precedence. A massive amount of data from multiple components might be generated and gathered to be stored and analyzed locally or remotely (i.e., cloud servers) to manage the flow of information.

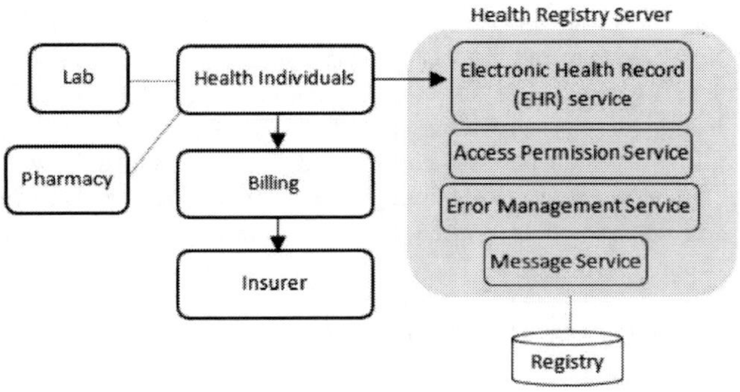


Figure. 1. Healthcare Information Flow

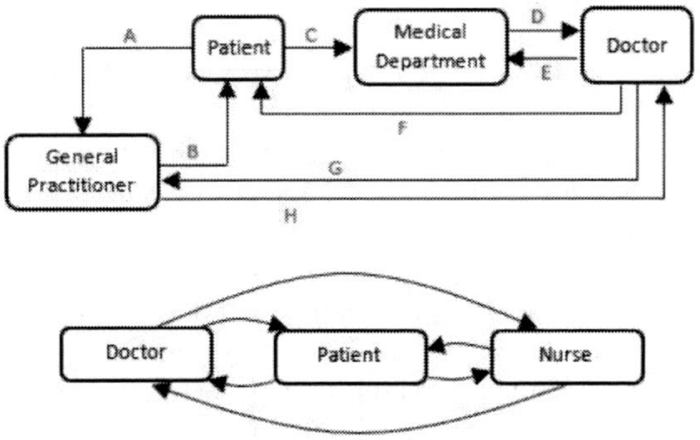


Figure. 2. Possible Processes Between The Communicated Individuals

In such a system, the information shared between participants might be subjective to several perspectives, such as:

- Organizational and managerial perspective: organizational boundaries, system participants' experience, variety of resources, trust between participants within and across organizations.

- Technological perspective: heterogeneous resources, multiple software, various security techniques.
- Policy perspective: the attempts to review policy processes in system workflow are challenging in resource constraint settings with less accessibility of policy, variety of policies, different policy versions, multiple policy path dependencies, and fragmentation of health sectors.
- Political perspective: routine, legislation, and information authority could limit health services offered by some providers.

The poor share of information could have consequences at different system levels. For example, at the computational level, the system's components resources (node's CPU, memory, network) could be saturated for several reasons such as security breaches, lack of resources, or process delay. Information exchange delays might happen in a healthcare system at the administrative level due to policy updates or lack of information. A healthcare system could be bounded by having many different policies that shape its system performance at the communication level. For example, a policy might prohibit general practitioners from obtaining a medical record directly from the records of a department without the permission of a hospital clinician. Electronic health records make it possible for healthcare individuals to share essential information at the security level. The share and open nature of interconnected health records across various organizational structures might allow access to sensitive medical data. For instance, medical staff could override restrictions to access sensitive data that were restricted in normal conditions when critical situations occur.

## IV. The Challenges of Healthcare Policy Change and Information Flow Management

Due to the distributed nature and dependencies between components of healthcare system clusters, safe interpretation of the information exchanged between such heterogeneous organizations, and healthcare systems might face several challenges such as changes in regulation, lack of functional integration, lack of support for delivery of information across system components, leak of sensitive information, and lack of interoperability among healthcare system components. In such a case, policy violations might happen and lead to performance degradation.

The Followings represent the challenges of healthcare and digital healthcare systems related to policy management, policy propagation, information flow, policy change, healthcare data storage, and information sharing focusing on monitoring, detecting, and recovering information flow.

### A. *Policy Change and Information Flow*

A healthcare policy defines the way information moves throughout a system. The policy is designed to preserve the confidentiality of data or integrity of data to prevent information from flowing to nonauthorized healthcare participants. However, healthcare applications involve dynamic requirements, which motivates the development of various kinds of dynamic policies as follows.

*1) Domestic Regulations:* Healthcare policy involves the creation and implementation of laws, rules, and regulations for managing nation's healthcare system. Some factors could condition information flow when a policy change propagated to the system, such as domestic politics, including political constraints, ideological preferences of politicians, policy participants, and policy entrepreneurs. One way to improve healthcare is through change policy, which involves decision-makers and stakeholders from different institutions; however, it is difficult to change policies because of dependency paths between different institutions. Healthcare agencies might be slow, bureaucratic, and inefficient. Once a country or a region has started a policy change, the cost of a reversal is very high. Moreover, in some areas of regulation, the division of authority is not clearly outlined, a drawback that has sometimes led to chaotic results. Any regulation that affects the healthcare system also affects the quality of healthcare. Thus, monitoring transparency in regulation is required to ensure quality-assured, safe, and effective healthcare services. A simple approach could be to provide interval forecasts of the policy change/update impacts of legislation considering cost and time.

*2) Multiplicity of Policies:* A healthcare system has several policies composed of different rules with their constraints, resources, parameters, and strategies. A change within policies could influence the internal and external system processes. Every process in a system might be executed in a given run, making correlations among them nontrivial. For different policies, the system might be re-analyzed in case of a policy change to track dependencies between the system's processes at run-time and at the same time to use the collected set of dependencies to check its processes against possible correlations for potential indirect information flow. Healthcare organizations find it difficult to effectively manage information flows within and across healthcare systems as it flows within and across diverse organizations. Thus, the change of policy results in many problems such as inappropriate decision making, less care management, poor quality assessment, ineffective planning, increase in medical errors, high cost, and a decline in the quality of patients' care.

*3) Resource Utilization:* When a policy change occurs, a system should cooperate correctly with the existing infrastructure to cope with the new applied rules and constraints and avoid rewriting all existing code to account for information flow constraints. Here, information flow congestion could mainly happen due to the intensive use of resources. Even though the policy rules are pre-specified, certain policy alterations could sometimes cause intensive resource use. Several static [21] and dynamic [22] analysis techniques are used to track the change in information flow and policy change; however, the literature studies neither provided run-time dependency tracking nor addressed indirect flows [23]–[26]. Moreover, current systems are not designed with an automated mechanism to dynamically adapt to the change in information flow policies [19][27].

The system should be able to predict future health services, healthcare needs, and rates of utilization of services and resources based on a foreknowledge acquired through a systematic process. The main idea is to monitor the current state of system resources and deal with a workload based on the current availability of the system cluster's capacity to satisfy their optimal performance parameters without causing performance degradation. A good understanding of the healthcare system requires analyzing the current and historical data to predict future needs accurately. Thus, reliable health forecasting could create alerts for the management of information flow and scientifically allocate resources to reduce the costs of supplies and resource redundancy.

*4) Leak of Information:* Another challenge for changing policy is that it might allow information to flow from personal data to public observers. Electronic IDs and tokens are used to prevent leaks and theft of information. However, current works [28] are not practical enough to handle the runtime policy change, or their method was inconsistent with the decentralization nature. Some works [29]–[31] rely on computationally heavy modular exponential operations or elliptic curve point multiplication operations. However, those works are very complex as the number of involved calculations is enormous, the structure of the data stored in the blockchain is difficult to query within a blockchain, limiting data usage, and the problem of efficiency remains to be solved. Noninterference is utilized to keep the flow of information within the authorized parties; however, many applications permit such downward flows according to their defined security policy, and some valid programs are rejected as insecure. Thus, specifying the kind of policies that includes downgrading, determining the nature of a downgrading mechanism, and the kinds of security guarantee are challenging because current approaches are too restrictive and challenging to enforce [32].

*5) Lack of Interaction:* Healthcare participants faced interaction difficulties due to situational and organizational factors. Lack of interaction leads to miscommunication and duplicate work. The cost of upgrading the system with the changed policy could be very high, which makes justifying the return of investment a challenging task. The use of cloud-based communication platforms like Unified Communications as a Service (UCaaS) is highly effective in breaking down communication barriers. However, the quality and availability of UCaaS services are tied to an organization's unique use case and specific set of business demands. When degradations in the quality of services occur, it is difficult to pinpoint the source of the problems in such a complex environment, making it challenging to ensure the quality of the healthcare system participants experience. A key to successful information flow while policy change is the seamless interaction between healthcare participants across and within the system. It is required to understand and control the overall effect of governing policies on the system behavior. The system should be self adaptable to the changes in policy and adjust its flow accordingly so the system's participants could receive the information smoothly. The correctness of the adaptation process should consider system stability, service availability, and resource capacity. Strategies should also be in place to help build the empathetic, relationship-building skills required to understand the patient's perspective.

### *B. Poor Sharing of Health Information*

The Healthcare system requires collaborative efforts from diverse healthcare providers and institutions to provide care for patients; however, since a patient is no more a passive recipient of care, the provision of high-quality services becomes indispensable due to:

*1) Tangled System Structure:* Patients' information is stored in diverse registries and repositories across different healthcare organizations to simplify authorized access to healthcare participants within the same institution. In such a system, many participants are involved in delivering care services, each having their interests, concerns, rules, and constraints. Some works [33] only handled hierarchical system structures without dealing with complex interrelationships between system variables. Network analysis techniques such as the Bayesian network [34] enable the estimation of the probabilities of system states and their covariates. Graph pruning is utilized to remove weak dependencies in the system structure [35]. However, it is challenging to choose prior probabilities and their appropriate probability distributions in such a dynamic system, especially in the presence of missing data.

Here, the conceptual framework must be set up clearly to identify the hierarchical structure. However, it is difficult to establish a relationship among all care entities as the system has a hierarchical dependent nature. Failure to account for the hierarchical structure could result in models that lead to unclear and misleading interpretations of the relationships under investigation, which results in additional costs in terms of resources, protocol complexity, and performance.

Thus, fragmentation and duplication of patients' information might happen, which impedes the ability of diverse healthcare practitioners to share data and gain access to patients' vital information. Accessing sensitive patients' information is complex and time-consuming for health care participants as it results in high transaction latency and low transaction throughput. Also, due to the dynamic nature of information flow and the change within policy rules, the quality of information is variable and unreliable. Hence, authorization and privacy should be dynamically adapted to the new policy rules to share patient data among different institutions within the system and give patients access to their records.

*2) Information Inconsistency and Incomplete:* Healthcare providers are usually presented with incomplete and inconsistent information during care. In such a case, information storage and retrieval problems might happen due to an information gap between medical participants and patients. The incomplete exchange of healthcare information during care transitions might cause ineffective care and additional healthcare spending [36]. Here, rollback of process or correction is one of the biggest challenges as it may lead to the waste of many months or years of research. Detection mechanisms, as the work in [37], could be utilized to identify inconsistency in information. However, some mechanisms are not flexible to be used in large-scale distributed systems, and they lack the support for handling dynamic policy change.

*3) Big Data Analytics in Healthcare:* Another challenge is utilizing different data formats. Medical data is not uniform. Imaging data comes in all different formats, for example, X-Rays will store differently from MRIs. Moreover, general hospital images are different from specialist hospitals that leverage more complex technology to ascertain more intricate images. Here, it is not easy to ensure data captured is clean, complete, accurate, and formatted correctly for use in multiple systems. Thus, EHRs are less interoperable and not easily deployable as there is no standard data format in the healthcare industry. EHRs require efficient automated or manual updates because medical data change minute-by-minute — this poses challenges in determining how to update quickly without end-user downtime and without slowing the system processes. A series of templates should be developed and included in the system's architecture to create an EHR standard to support an interoperable system. Thus, healthcare systems need integration of healthcare standards data models, such as Health Level Seven (HL7), Fast Healthcare Interoperability Resources (FHIR), or open EHR archetype along with terminologies like SNOMED-CT, to provide timely access to healthcare information, reduce the administrative burden from providers, offer one common integrated system across all care settings, and to seamlessly share data across multiple settings (e.g., OpenHIE and IHE).

Different studies [13][20][38] explore approaches to solving interoperability problems. However, there are difficulties in adopting health standards and tools for adequate data representation (ontologies, databases, clinical models) that ensure healthcare professionals efficiently manage the data, such as:

- Unstructured text fields are not readable to the machines,
- The combination of text fields with health standards was almost unanimous due to a low variety of semantic healthcare web technology,
- The storage solution is related to the type of adopted healthcare standard,
- The usage of use of ontologies impacts the choice of storage solution since querying the ontological structure is language-dependent,
- Data are sometimes not compatible for exchanging information,
- Health and medical concepts and terms used across the organization do not preserve their meaning when externally shared,
- Healthcare applications that use common EHR standards for data sharing might not ensure confidentiality and privacy of patient's sensitive health records that are shared in closed and open networks.

Until information sharing is addressed adequately, poor communication often causes cancellation of procedures, loss of revenue, and inefficient resource utilization. Thus, healthcare organizations need comprehensive auditing and tracking features to guarantee compliance. Organizations need the ability to perform patient record matching, where error and duplicate rates are monitored, and any access to patients' records is detected. Hence, health participants and providers could share sensitive patient data securely within and across systems with the proper information protection strategies. Sharing healthcare information requires different levels of integration within and across organizations. The need for securing EHR differs within the same organization's participants. A dynamic and robust technique should be designed to permit the secured sharing of sensitive health data in the disparate interoperable healthcare domain. According to HealthIT [39], healthcare participants should be provided with self-adaptable services that allow them: to search for and access electronic health information within their workflow, seamlessly integrate electronic data from inside and outside the healthcare system, and set preferences and control how they can share the electronic records, with whom, and for what purposes.

### C. Healthcare Registry Management

One of the main information flow between system components is information pathways between hospital care services and specialist consultation. Along the path, different healthcare providers are involved and use different electronic systems, which are also used for different purposes within a single organization. For instance, the municipality uses one electronic medical record system for documentation, while hospitals and primary healthcare services use another. In such cases, various data types are used with no common format for holding it commonly. The information flow among healthcare participants, other organizations, and institutions needs to be organized and visualized so that information can be accessed at any time and in any place.

Due to the lack of integration between various system components, health participants and patients might have limited access and control over the collected data within a registry network. In case registries have been developed and data are collected, it might be challenging to modify the established data collection procedures. A patient's informed consent should allow data to be shared within a registry network. However, if there are established registries at a country or region level, network models might be more efficient in terms of costs and time. These registries are time-consuming and resource-intensive to establish, particularly if large numbers of patients or long-term follow-up data are needed. Moreover, registries using network models might face some additional challenges related to governance, data harmonization, data sharing, and change management. Some registries incorporate information from other data sources, such as electronic health record data.

For multinational registries, linkage and access to other data sources might be challenging due to varying requirements and availability for accessing such sources [14]. Data records that are available in certain countries might be restricted to particular regions or localities and might be difficult to link with individual patients.

### D. Management of Policy Propagation in Multi-Cluster System

Healthcare policy poses rules and complex legal to protect the health of individuals.

Policy rules are associated with policies to data flow steps. Some rules and descriptions are needed to propagate policy, such as a description of policies attached to data sources, a description of data flow (the actions performed on the data), and policy propagation rules (which actions do propagate a given policy). This activity results in many numbers of rules to be stored and managed.

As policies are associated with licenses, a suitable mechanism is required to check the compatibility of licenses and to validate constraints attached to components in a multi-cluster system. Data Hubs are used to collect a large variety of data sources and process them to implement the workflow that connects data in their sources to applications that might exploit these data. These systems create new challenges in terms of the volume of data to be stored and require novel processing techniques such as stream-based analysis to govern data. Assessing what policies propagate from the licenses associated with the data sources to the output of a given data-intensive process is a significant problem. Policy rules could be compressed using an ontology of the possible relations between data entities [22].

However, the ontology matching problem could arise due to finding the semantic mappings between entities of two given ontologies. The coherency check algorithm allows effective reasoning with a compressed rule base but assessing policy rules' coherence demands a partnership between participants of different healthcare sectors and requires cache coherency awareness.

A generic graph matching algorithm is used to match and convert schema into directed labeled graphs and then uses fix-point computation to determine correspondent entities (nodes) in the graphs. However, the algorithm does not function accurately:

- If there is no label for the graph arcs,
- In case the labels are almost identical, or
- If there are a few levels and most of the relations are associated with the concepts at the top of the graph.

Besides that, several methods [26] are used to enforce information flow policy propagation. Run-time mechanisms that tag data with information flow labels have been employed at operating system and programming language levels. Static analyses have also been developed to ensure that information flow within the system complies with policy rules. The problem of false positives and negatives is less in the case of dynamic analysis because they analyze by running the test cases. However, dynamic analyses are inaccurate as they cannot observe all execution paths. They require a large number of test cases to ensure a certain confidence level in detecting violations and to cover all information pathways. Thus, to guarantee noninterference, the techniques either terminate executions that might release sensitive information or ignore updates that might leak information.

Another method to enforce information flow policy propagation is through a security type system, which enforces security properties within a system application. Hence, the system allows the flow of information for the changed policy if it is type-checked and contains no improper information flow.

Nevertheless, such methods are too strict for use in most real-world applications, such as non-interference policy. Thus, several approaches are proposed to control various policy releases, such as information declassification and formal modeling, so an active attacker might not manipulate the system to learn more secrets than what passive attackers already know. However, the performance of evaluating those approaches is complex in the case of analytical methods or discrete event simulation. Regardless of their dependability on specific commercial-off-the-shelf simulation packages, more invasive revisions of their model are needed, especially with policy changes. Controlling policy change and its versions should contemplate its relations with system configurations and tangled clusters, considering the maximum capacity of the cluster and cost.

### E. Information Flow and Policy Change - Monitor

Monitoring information flow and policy change is a way to consider the system state and the execution paths. Here, we should consider the system's current state and the executed paths and non-executed paths to monitor the policy change and ensure that sensitive data will not be revealed to unauthorized parties.

Data label tags and semantic rules could be used to monitor information flow for sequential processes and observe the application's inputs and current values in variables. Each application input receives a tag, which reflects its security level and the current value of the variable. Here, an evaluation of rules is applied to return a value, and a tag is created to reflect both all the previous information flow and the generated one by the evaluation [23].

However, monitoring the information flow of changed policy and its prorogation is challenging in a multi-cluster environment because of the diversity of data structures, various data type formats, dependencies among system components, and the precedence between system components. Here, a directional graph could be used to represent each subject as a node, and the flow of information is represented as a directed edge. The graph could be further refined by using an information flow vector to divide the complex directional graph into sub-layers and quantifying the flow of information [25][26]. An authentication mechanism could be used to compare authenticated credentials to a set of known credentials to determine the access of authorized applications. Nevertheless, massive amounts of data and rules might be generated, aggregated, processed, and stored. Collecting a large variety of data sources, and processing them to implement a workflow that connects data, creates new challenges in terms of the volume of data to be streamed, analyzed, managed, and stored. Such data could be associated with new rules and constraints that require policy rules to be adaptable to reflect and propagate the new modifications to the system. Hence, it is challenging to develop a monitoring technique that coordinates various monitor instances running locally on the edge gateways and optimizes the communication cost when processing the data while minimizing the overhead of CPU, memory, and network.

Thus, coherency check algorithms, properties matching, static analysis or tag checking algorithms allow practical reasoning about the change of policy rules. However, several challenges have arisen [21][22], such as identifying policy properties to be monitored at different system levels. Changing or updating a piece of data or a policy rule might alter the hierarchical dependency between system components and might affect the flow of information within the system. The situation becomes complicated when hundreds of global regulations, federals, states, and region-specific mandates with too many updates and versions of policies. Every time a policy is impacted by a change in regulation, it goes through a cycle of updates, reviews, approvals, communication, and attestations.

Hence, when changes happen in a policy, the modifications affect the system's participants. Unless new policies become established, organizational performance might be negatively affected as participants become accustomed to new ways of performing tasks or different expectations for personal behavior. An overhead might happen to a system components due to strangulation of the information flow in processes. In such a case, the performance of services, applications, nodes, and communication networks vary significantly depending on runtime variations in running conditions such as availability of resources and the network connection quality between different application components distributed over the Internet. Thus, several factors should be considered during policy change/update, such as hierarchical dependency between system components, the flow of information in processes, the impact of policy changes on the system's properties, and the necessity of the change/update and its relevance.

Monitoring policy at every system's stage helps identify and address problems in information flow pathways. Hence, a systematic and consistent solution is needed to integrate policy change management to be forwarded faster and mitigate compliance risks.

### *F. Information Flow and Policy Change - Detection*

Dynamic detection is a technique that leverages metadata tags to track the information flow and policy change among different entities.

Information flow policies define the authorized paths throughout a system. These policies are designed to preserve confidentiality and integrity of data by associating labels and rules to represent a security class with information and entities containing that information and enforce some rules about the conditions under which data could flow throughout the system.

Several techniques of information flow are proposed to detect policy changes and to analyze them for signs of possible incidents such as violations of security policies or non-acceptable use of policies [21][24][40]. Static and dynamic analysis techniques are used for verifying a program's compliance with information flow policy. However, static analyses are less precise than dynamic analyses, as they consider only the executable flow paths, generate many false positives, and provide no runtime dependency tracking mechanism. Static and dynamic analyses are combined to reduce the false positives and minimize uncertainty within both information flow and policy propagation that could arise due to lack of information, growing scale, and complexity of the data. Thus, machine learning techniques are used to build detection models that characterize the activity of the system's components at runtime [41][42]; however, they require high computational power, and they need to be trained from the beginning if new data arrive.

Hence, examining logs, identifying new rules and labels, and checking tags during runtime are used to simplify detecting illegal information flow and determine policy violations. However, applying those mechanisms to data flow might be a complex process as they often require merging data from disparate sources. Such data merging might cause potential disclosure of sensitive information, data redundancy, lack of interoperability, shortfall of data sharing, workflow interruption, and barriers of interdependency between system's processes. Thus, to tackle those challenges, some factors should be considered, such as: identifying properties that effectively could be used to detect the policy change/update. Here, it is difficult to detect the change in policies because of the dependency between the system's clusters and its components and their impact on the system's constraints and resources. Moreover, the eligibility of policy's change/update during runtime should be checked to prevent illegal information flow to the system's components, as illegal flow might occur even if every access request is authorized.

Because of the complexity of policy's rules change/update, and the varying degrees of security levels between clusters, it is challenging to make these systems secure as security is a crosscutting requirement scattered over distributed clusters. A mismatch or local vulnerability between security mechanisms adopted at different clusters might impact the flow of information between clusters and cause an overall system's performance degradation. Thus, detecting policy change in a distributed system requires keeping track of all processes, rules, and resources and applying an authentication mechanism to each cluster's components.

Here, changing policy rules might allow unauthorized access. Users acquire their necessary permissions by being assigned membership to suitable roles; this might significantly reduce the system's overhead since users with similar access requirements are grouped into the same set of roles, and the requisite permissions are included in those roles. However, in such a system, policy's rules might be restricted to the share of resources, the hierarchical directed or in-directed relationship, and the precedence between multiple distributed clusters. Hence, tags are used to indicate different security policies with different instruction types; however, utilizing tags might cause a waste of tag storage as

- Tags' size grows with the incremental flow of data,
- Not all data are involved in computing, and
- Many tags might not be used at runtime.

Moreover, different tag propagation and tag check rules might be customized according to corresponding instruction types [24][43]. Here, performance overheads might occur, and the complexity of tag storage might be increased due to the utilization of extra detecting operations. That, in turn, introduced high hardware overhead, high false positives/negatives values, and a lack of flexibility for specific types of policy reconfiguration for different program contexts (e.g., security policy).

Other works, such as in [44], utilized prespecified heuristic rules, which require experts to be defined based on the historical knowledge of system behavior.

Thus, a dynamic approach for managing information flow could be applied to cope with policy changes and eliminate repeated and inefficient flows.

### G. *Information Flow and Policy Change - Recovery*

Healthcare systems recovery is defined as *"the rebuilding, restoration and improvement of the healthcare system's components and its core functions, in alignment with the principles of building sustainable development"* [12].

Healthcare system processes are performed sequentially or simultaneously by various participants within and across the system, and they are organized through several tasks in a workflow. The workflow might be subject to vulnerabilities such as malicious attacks. A malicious attacker might create an illegal task or corrupt a task in the workflow. Such malicious tasks would possibly corrupt data items accessed by some benevolent tasks, or it might trigger other workflow tasks due to dependencies and precedence between them. Moreover, tasks that depend upon malicious tasks might be corrupted and might affect the flow of information within the system.

Some literature studies investigated the recovery from malicious attacks [45]. In such systems, a transaction executed by a malicious attacker might corrupt other transactions in the workflow. Techniques and algorithms are used for assessing and recovering that damage, such as parsing a database log to check which transactions are affected by malicious transactions and undoing/redoing the affected transactions. Store the dependent transactions in separate structures is also applied to preserve a log from being traversed for damage assessment and to be used later for repair. A backup service is utilized as a way of recovery to allow restoring the system's state after a severe attack.

However, such techniques for recovering damage are not adequate for workflows as transactions in a database are independent entities. Hence, a workflow classification is utilized to classify workflow into documents, processes, and system workflows to identify potential types of attacks and to restore the most recent consistent process state after a failure or rollback of inconsistent execution of interrupted tasks [40]. Another recovery way is to resume the execution of the process from the closest consistent point where the attack occurred [46].

A recovery action could be taken on a particular occurrence with an immediate response at the desired state or a set of states to apply a recovery mechanism. Here, the system should calculate its current state and capabilities depending on a set of actions to transition from the current state to the desired one. A resource controller could help assign spare resources to the requesting cluster's components. Once the resources are reserved, the cluster regulates the allocation and reallocation of the assigned resources considering their utility optimization and cost reduction.

Alternatively, a recovery could be made by removing a compromised task from the process, restoring the process to a normal state, cleaning up corrupted data in data memory, and releasing the resources taken by the compromised task.

Hence, for deciding which recovery action to take when a policy change is detected, a mechanism should automatically determine one or more recovery plans based on the type of detected change, considering the control and information flow among tasks across the impacted components in the system. Here, constraints might be applied to choose an alternative recovery action if the immediate previous action failed. In contrast to ranked actions, the execution order is computed dynamically at runtime; however, there are some challenges, such as defining a domain-specific language to describe the capabilities of each component in the system in terms of actions, roles, responsibilities, and information pathways. In such a case, high overhead during runtime might happen due to various constraints on the flow of information across distributed clusters.

Moreover, a deep understanding of the underlying infrastructure is required to develop a recovery-based strategy for determining an optimal recovery plan and recommending a set of successful recovery actions for a given violation. The strategy should consider the system's available resources, allowable information pathways, and regulatory and business constraints (e.g., computational budget). An executable model of a recovery plan could be designed to carry out the recovery in a distributed and coordinated way across various components in the system. Here, the system's resources, status, capabilities, and dependency between components should be considered to check the extent of damage after a failure.

Furthermore, suppose all system policy parties' response ends, and the recovery phase begins. In that case, the consensus of the system's parties might take some time before the recovery transition mechanism and structures are formalized, even if all parties agree that the transition phase has begun.

Ensuring resilient and responsive healthcare systems are vital to achieving the objective of Universal Health Coverage (UHC) Vision 2023, and for advancing progress on the Sustainable Development Goals (SDGs). Hence, creating effective health systems requires well planned and well-implemented recovery strategy.

## V. Healthcare Opportunities and Future Research Directions

The Healthcare system is a dynamic environment with significant opportunities such as cost concerns, service quality, uncertainty in new technology, and complexity due to diversity of tasks, diversity of care pathways, lack of sharing, the vulnerability of patients, dependency, and relationships between system's components. The healthcare system aims to improve, secure, and accelerate care services to participants. The following explains the potential opportunities and future research directions in healthcare systems, mainly focusing on information flow and policy change share, security, monitoring, detection, and recovery.

### A. Cost Reduction

Several factors could aid in reducing healthcare costs, such as utilizing information technology investments in the healthcare industry to increase profitability, quality of products, and services.

Providing online healthcare services decreases the processing costs of many activities compared with manual handling operations. Such a way might reduce the number of inefficient processes by allowing data sharing across multiple healthcare sectors, pooling the skills and capacities of healthcare participants for problem-solving, contributing to the elimination of mistakes from manual procedures, and reducing the required time for transactions.

Moreover, the cost of components, which are needed to support capabilities such as sensing, tracking, and controlling mechanisms, needs to be relatively inexpensive through utilizing cost reduction strategies (e.g., service standardization, performance tuning). Such strategies could prioritize patients' health while examining opportunities to lower overall costs and increase patient satisfaction.

However, the adaptability of the healthcare system might complicate attaining the goal of reducing the overall cost, as it is prone to inefficiencies such as unnecessary care, waste in healthcare, unwarranted clinical practice variation, administrative burden, and fraud. Thus, cost-containment policies might be applied to target all aspects of the healthcare system, such as prices, volumes, supply, demand, and market processes. An effective financing system should estimate a potential expenditure based on the volume of data and costs and might use the estimation to change the number of funding sources to meet budget constraints across different hospitals.

One of the cost reduction methods in the healthcare system is reassessing the organization's healthcare planning processes to improve accountability and agility of the overall healthcare participants and system's processes. Another way is to use an advanced cost accounting tool (e.g., Activity-based cost analyses, marginal cost analysis, minimum pricing analysis) to drive a deeper understanding of the system's targets and achieve them. Hence, an organization should invest in the healthcare processes to educate healthcare leaders on the usage of advanced cost accounting tools and understand the data.

The healthcare delivery system needs to automate both the connection of healthcare devices and the data centralization to reduce the cost of system management and patient care and to optimize the healthcare system's processes. An autonomous healthcare system could drastically reduce the system's cost management, decrease patient care costs, and improve the system's performance through reallocating resources according to the organization's needs.

### B. Healthcare System Services Quality

The quality of healthcare system service affects the satisfaction of healthcare individuals and could increase the likelihood of desired health outcomes. It consists of technical (type of delivered service to a patient) and functional components (service delivery process). These components contribute to healthcare quality and affect the success of its services through (1) providing direct and remote access to health and medical records. (2) Enable automatic data sorting to enhance the generation of information. (3) Reduce medical errors in healthcare services. (4) Analyze causes of a system failure. (5) Effectively communicate and collaborate with other health professionals or institutions. (6) Regular monitoring of the services based on importance and priority. (7) Avoid the suspension of healthcare services. (8) Support standardized treatment policies and protocols that minimize errors. (9) Develop a financing mechanism that supports continuous quality improvement. (10) Enable prevention, detection, and response to health security threats.

To increase the outcomes of healthcare and to minimize resource waste, the quality of healthcare system services should include characteristics such as availability, compatibility, performance, interoperability, accessibility, privacy, confidentiality, accuracy, reliability, and comprehensiveness [15]. The system should allow continuous, convenient, timely access to care services, compliance with clinical practice guidelines, and support continuous monitoring of patient conditions. These lead to accurate and comprehensive patient medical records, which increase the efficiency of diagnosis and treatment services.

Moreover, the ability to exchange and share records across different departments and organizations leads to cost efficiency, effective patient treatment, elimination of redundancy, enhancement of doctor-patient relationships, and enables authorized centralized care coordination to provide access to high-quality people-centered health services.

The quality of healthcare services mainly depends on participants' knowledge and technical skills. Some barriers impact the quality of the healthcare system, such as centralization, bureaucracy, and hierarchical dependency among organization and institution sectors. Such barriers might cause delays in the provision of healthcare services and might lead to a negative perception of the provided service quality in case it is unnecessary by participants.

Thus, the healthcare system should focus on: the design of participants-oriented-service processes, tracking whether the system's policies and processes are being met, and creating collaborations within and beyond the system. In such case, relational analysis (e.g., grey relational analysis, analytic hierarchy process) [47] could be utilized to improve the quality of healthcare system service delivery and to analyze the relations between system's services.

Furthermore, the quality of healthcare system services could be impacted by a service failure such as a service breach that might affect the service's availability and reliability. The severity of service failure and the occurrence or frequency of failure should be measured to manage and recover the failure according to the health quality and standards (e.g., CAHPS 6.P[48], Six Sigma Healthcare [49]).

Moreover, because of the heterogeneity inherent in services, different participants within the same organization might experience various instances of service failure and its recovery. However, if the service failure is detected early, the system's reliability and participants' satisfaction would be increased. Improving quality of health services requires good communication and collaboration among healthcare providers to provide effective and efficient services and promote shared responsibility to deliver the highest-quality care.

### C. Uncertainty in Internet of Healthcare Things (IoHTs)

IoT devices offer opportunities for healthcare provider organizations through providing remote tracking and monitoring to reduce healthcare costs, optimize resources, and provide accurate data collection.

As the number of connected IoT devices grows, the amount of generated data increases and becomes more complex, which requires a mechanism to analyze data and utilize repositories that hold all volumes of information. Thus, healthcare organizations find it challenging to adopt IoT because of:

- The lack of standards and security practices,
- The challenges of integrating data from IoT applications into legacy systems,
- Inadequate privacy regulations,
- The use of expired infrastructure,
- Lack of consensus regarding IoT protocols,
- The high cost of implementing IoT technologies,
- Diversity in devices calculation and communication capabilities, and
- resources constraints.

Several factors might influence the occurrence of uncertainty in IoT, such as:

*1) Security and Privacy:* Confidentiality and privacy are important concerns in healthcare. Low security and misconfigured device and network settings could affect patients' privacy and their data. The use of various providers mandated to submit confidential data to law enforcement agencies, this could affect the adoption and use of the technology. The networks that transmit data are often highly heterogeneous and are frequently managed by third parties, which makes the protection of security and privacy and governance of this data even more challenging. Moreover, an organization could identify risks associated with IoT devices and should preauthorize the security team to help remove vulnerable devices from the network during attacks. It could be aware of all assets that can impact the security of the healthcare IoT network such as limit access to the IoT network, separate network segment for IoT devices, encrypt the network, protect the data, ensure and manage the communication of authorized devices.

*2) Hardware and Software Failures:* Hardware and software failures might put healthcare tasks at risk, as a delay might occur to one or more processes, which might propagate to the entire workflow. As a result, multiple devices might not work well together, or the functionality of devices produced by different manufacturers might have varying characteristics. In such a case, a security breach in IoT might occur and leak personal information to unauthorized participants or a device malfunction. Furthermore, with different hardware solutions, the software has to be timely updated to a safely stay at its latest version and allows the aggregation of data from various devices. Such constant updates require lots of effort and might spawn many technical issues. Hence, policy should strengthen current requirements for data exchange among electronic health records, the emerging IoT devices, and solutions. It might maintain an inventory of all healthcare systems connected to the network so that organization could quickly identify, and address risks associated with IoT devices.

*3) Devices Resources Constrains:* In healthcare, many devices are connected, producing a massive amount of data and information, which might affect the computational and processing capabilities of devices. Offloading could be utilized to partition and execute tasks between devices and edge nodes to minimize energy consumption [50]. A mathematical programming-based framework is utilized to allocate tasks while satisfying operational constraints optimally [51]. However, those ways focus on optimizing limited system parameters (e.g., processing capacity and network bandwidth, and couldn't handle the dynamic nature of policy change.

*4) Interoperability and Devices Heterogeneity:* Many devices now have sensors to collect data and often communicate with a server in their language. Each manufacturer has its proprietary protocol, which means sensors made by different manufacturers cannot necessarily communicate with each other. Device management requires directories of devices' functionality, protocols, terminologies, and standards compliance. Interoperability is a significant challenge in creating medical devices that easily connect with other devices and sensors to health providers' electronic medical record systems. However, it takes time to corroborate such a massive amount of data with the different terminologies and standards on every system and might even yield inaccurate results. Moreover, current health standards are challenging to use due to the lack of adequate code for electronic medical records and because they describe fixed timing of examination results, not being a sequence of patient episodes. Thus, migrating to more interoperable technical solutions should ensure continuous and uninterrupted services and provide universal guidelines on the consistency and agreement of data in a format, queries, and synchronization. The solution should support workflow consistency in how technology helps decision support, clinical guidelines, rules, and user interfaces. It should provide an approach that deals with different device calculation and communication capabilities to share, understand, interpret and use data without ambiguity. The solution should use low-cost interfaces [52] and open APIs (such as HL7 and FHIR) to solve the usability and cost issues as they distinctly define security mandates and transmission protocols which lower the risk of errors and have real-time ability to share and access data.

### D. *Enhance Healthcare Services Security - Blockchain*

The healthcare system involves many processes such as emergency department operations, managing finances, and patient transfers to different facilities. Such processes constitute one or more workflows, which might involve repetitive tasks related to one or more aspects (e.g., patient transfers can be plotted out as a series of conditional tasks). To provide better internal controls that minimize risks, eliminate workflow cycles, and reduce overhead, the healthcare system focuses on protecting its processes and services against unauthorized access, hardware theft, data manipulation, and common threats and exposures. Thus, rules and principles are enforced to regulate the access and transmission of data to healthcare participants and providers. Here, blockchain techniques (e.g., public, private, hybrid, and consortium) appeared to provide a secure, authentic, and transparent distributed technology that could integrate the healthcare system's services from multiple nodes in the blockchain network to enhance their computation, storage, and transactions processing. Furthermore, it provides:

- Better healthcare data-sharing,
- Assists in the diverse use cases of healthcare,
- Trace data shared within and across a business network to provide well-controlled privileges to healthcare participants.

To safely exchange healthcare information between authorized participants, blockchain-oriented platforms [53], and blockchain models [54] could be used to evaluate healthcare data sharing requirements in different sources, validate data accuracy and patient engagement, and improve the dissemination of accumulated information in a secure, interoperable environment. Other models [55][56] are used to evaluate the performance of new patient block components and system configurations, improve data accessibility between healthcare providers, and provide a secure runtime monitoring system. Such a system enables healthcare participants to track the healthcare status of their patients remotely while safely maintaining the latest history of patients. Here, authorized participants only could view patients' identities within the authorized network. A hyperledger platform [55] is one of the techniques that could be used to provide patient-controlled healthcare data management.

To protect patients' sensitive data from being tampered with and to eliminate the cascading of malicious behavior to the overall system, digital signature introduces another level of authenticity. It uses a cryptographic operation that binds the signature and the signed data by adopting the idea of the Public Key Infrastructure (PKI) feature. A unique digital signature is issued to lock the data and prevent any additional signatures, annotations, or data fill-ins.

Another way is to use smart-contract-enabled blockchains [56] like Ethereum to create digital currencies (tokens), which could remotely monitor the healthcare system, and securely identify, authenticate, and authorize system participants. Here, smart-contract might be utilized to create representations of existing health and medical records that are stored within individual system components. The contract might contain records of metadata ownership, permissions, data integrity, relationships, and state transition functions to carry out policies and enforce a set of rules regulating specific records access.

Hence, blockchain allows the distribution of EHRs among different health care sectors to manage patient-sensitive information through several nodes of an interconnected network. However, correctly arranging the gathered data and defining their dependency and precedence in a blockchain network are problems due to the existence of private transactions and the concept of cryptographic protocols that allow private calculation of encrypted transactions to be accessible into the blockchain. Thus, blockchain healthcare technologies could optimize system performance while retaining small processing and computation capabilities for data representation. Moreover, the most promising applications of blockchain in healthcare are for dynamic patient consent and identity management. However, to enhance the healthcare sector, the healthcare applications need to support big data scalability to deal with the massive amount of health data, cross-border health data, and their policies. As the volume of data and transactions increases, a mechanism is needed to minimize the delay of storing and processing massive data access transactions considering the incomplete and missing data provided.

### E. *Improve Healthcare System Recovery*

The ultimate goal of healthcare system recovery is to design a system that can respond to the dynamic demands, perform its functions effectively and sustainably, increase health systems resilience, and mitigate the risk of future healthcare policy change. Health system recovery is determined or influenced by the typology of emergency care. Hence, to understand the system's processes and its tasks, it is mandatory to understand the need for recovery and the emergency of recovery to develop a recovery plan. The recovery might be based on restoring the system to a specific point while investing in risk reduction and strengthening preparedness for future hazards.

To enhance the healthcare system, some critical elements for enhancing the system's recovery could include:

*1) Recovery Duration:* Some criteria could be used to assess the phase of recovery, such as urgency, sustainability, and cost-effectiveness of the tasks being carried out. Recovery duration is another criterion that could impact the recovery phase. Once the initial recovery has been carried out, other residual or outstanding activities could be integrated into subsequent standard planning mechanisms. The duration of the recovery phase could be a short-term or long-term recovery.

However, differences in the types and evolution of emergencies might increase due to an overlap between the two phases and emergency response and recovery. Thus, dividing the recovery process into phases is essential to facilitate standardization and harmonization.

*2) Recovery Conflicts:* Conflicts are typically protracted, and there might be an assortment of health participants with different degrees of legitimacy from different organizations. When multiple participants are involved, recovery might be delayed, and efforts could be duplicated as recovery from conflicts passes through a long process of restoring the capacity of the health system's components and core public health functions. The process includes restoring the government and communities' capacity to rebuild and recover from crises and preventing relapses. Moreover, a system with various resources, capacities, and capabilities of its participants' should be supported by a government throughout each recovery phase to avoid recovery conflict within and across the healthcare organizations. Although it is critical to have a concerted health sector recovery mechanism led by a government to focus on funding and resources, a complete understanding of the process is necessary to distinguish between conflict and non-conflict-related emergencies.

*3) Up-To-Date Recovery Mechanism:* Hence, self-healing systems could be adopted to take corrective recovery actions and trigger an alert if a system does not satisfy constraints while identifying lessons learned from healthcare members to support future recovery. Thus, to meet recovery objectives, developing a recovery plan should support dynamic monitoring tools and evaluation mechanisms to measure the progress of the recovery process and its outcomes. The plan might identify funding sources for an early recovery plan and provide a recovery strategy with contingency plans for different scenarios. The development of a recovery strategy and plan should be based on the results of the recovery assessments. Thus, a recovery mechanism should adapt to the new integration principles to ensure that the recovery is aligned with national priorities across healthcare sectors to manage unavoidable trade-offs between short and long-term recovery priorities and economic and environmental policy goals. The recovery mechanism should ensure that the recovery responsibilities, including financial requirements and policy arrangements, are transferred to support the recovery phase. In conjunction with the recovery process, an assessment of post-recovery is needed to determine the overall recovery funding needs. Thus, adapting public financial management systems could support the government and provide international best recovery practices. The recovery mechanism should adapt to the new health sector recovery roles, responsibilities, and priorities, including those of the municipality, the district private health sector, and partners. The recovery action should consider the coordinated entities within and across the organizations, even with prior regulations, strategies, mechanisms, and platforms.

*4) Recovery Consistency:* A consistent recovery between healthcare system organizations and their participants helps meet their perspective needs. The coordination and definition of roles and responsibilities during recovery should be supported by national legislation or a memorandum of understanding before a recovery mechanism formalizes the roles or before the failure spreads to the whole system. The recovery mechanism should be confirmed with health sector partners to be used for consulting stakeholders on the draft recovery plan. The mechanism should provide a recovery guide for the use of stakeholders to identify priority health sector repairs with partners. The mechanism should allocate the cost and time of immediate recovers required for the system. During this period, a cost indemnity agreement should be designed and agreed upon within and across healthcare organizations to guarantee immediate recovery action.

*5) Seamless Communication and Collaboration:* As healthcare is a tangled system with multiple participants and protocols, data-sharing protocols and agreements should be developed by organizations that have access to and stewardship of required data. The recovery mechanism should consider the variety of data-sharing protocols and agreements to facilitate communications and collaboration and support real-time information flows between and among health organizations. The mechanism should ensure consistent and up-to-date repair to reduce duplication [16]. A gap evaluation analysis is required to assess system capacity and capability to meet recovery objectives and identify the differences between the current state of the system and the desired one. The mechanism should re-assess the initial needs and the recovery plan (inputs, outputs, results), modify the process as required by results, and disaggregate the indicators under observations. A recovery index of health organizations (e.g., WHO and PAHO recovery health index) could be used to assess the impact of recovery on the system and provide oversight over health and safety guidelines to repair the healthcare infrastructure. The recovery should be related to the system's architecture. Here, a systematic way could be developed to provide a sound basis for making objective decisions about design trade-offs and to enable accurate predictions about the system's capabilities and qualities free from bias and hidden assumptions.

### *F. Develop Dynamic Monitor for Healthcare Information Flow and Policy Change*

Continuous monitoring of the health system's services helps identify bottlenecks in information flow. Such bottlenecks could cause delays, overload, and blockages in workflow processing, leading to dissatisfied health participants, loss of revenue, time, and wasted resources. Continuous monitoring creates a vast bulk of data that requires an expert system to perform analysis and processing dynamically.

To monitor the change within policy and control its impact on information flow, the interaction and the communication mechanisms of persistent data, processes, and sub-process should be analyzed and presented in various visual formats and charts. Current monitoring tools cannot provide information about patient health status, and they do not visualize all the recorded data on the same platform [23][57][58].

Furthermore, determining which system's service is not satisfactorily performing could make better decisions about reconfiguring components, services, and resources to run processes more smoothly. Heterogeneous computing resources should be reconfigured and scaled according to the monitoring data in a distributed environment. Here, system components and services might contain some redundant processes and dependencies, which introduce noise in a workflow that adversely impacts the performance of inline and real-time system analysis. Thus, noise reduction at the process level could be used to reduce computational complexity and achieve better system performance.

Hence, monitoring could be carried out at different system levels to provide continuous information flow monitoring with the policy rules that might change over time and incorporate every change into the system without causing service interruption. A notification also might be triggered for any workflow processes that might conflict with given policy rules without re-analyzing the whole workflow. Thus, the monitor mechanism should:

- Distinguish the discrete processing stages within the processes.
- Describe the information flow mechanism through a system.
- Characterize the type of data items that flow through the processes.
- Identify violations within information flow and policy rules.
- Find locations for inserting data validation monitors, insert data collection points for later analysis, and distinguish between data dependencies and control dependencies.

A monitor system should consider a real-time wireless transmission connection between the monitoring tool and the monitored device to improve the monitoring accuracy. Monitoring properties associated with a potential risk prediction should be further analyzed to overcome challenges such as unpredicted faults, massive data streaming, and detection accuracy. The amount of collected data from monitored properties is large and intractable. Hence, it is required that the monitoring system distinguishes between health degradation and faulty measurements.

### G. Continuous Detection for Healthcare Information Flow and Policy Change

Detection is the process of comparing current normal processes with the observed ones to identify significant deviations. In such a case, policy properties (point-method) could be identified in the monitored data and the information flow, or a score (likelihood-ratio-method) could be associated with the arrived data to indicate how likely there is a change in the policy. In both methods, the main goal is to reduce detection delay.

Furthermore, label-method might be used to detect the policy change for each component in the system. However, in a dynamic multi-cluster environment, statistical methods do not provide accurate results due to the regular update of the system's processes and policy rules. They are ineffective in detecting real-time noise changes given limited information flow.

Here, multiple components could be existed within each cluster in a system, along with dependencies and precedence among them. In such a system, detecting the change in policy rules might be difficult because all the flow of information and processes among components should be understood according to the updated policy. Thus, continuous policy change within and across such a hierarchical system requires a complete understanding of policy rules and needs to trace executed and non-executed paths of information flow and processes. Hence, a detection mechanism might be trained with sufficient behavioral and temporal features, work on each sub-process, and decide whether a policy is changed to avoid policy conflict and prevent the system's workflow delay or bottleneck. According to [16][59], the detection mechanism should:

- Minimize the detection errors to provide an efficient system that can be managed remotely.
- Support low computational complexity and few amounts of memory to store training data.
- Be practical in health and medical applications to reduce redundancy within the monitored parameters.
- Classify different types of policy change based on the measured time series (e.g., regular time change, noisy time change, short time change, long time change).

Moreover, the detection mechanism should be self-adaptable to deal with dynamic changes within policies during runtime, identify relevant changes, and compare policies across and within a system. Once the change is detected, the mechanism might give a participant the ability to modify the information flow or policy rule during runtime and choose whether to continue, alternate, or abort the execution of a process. The detection should help spread the approved policy change to the rest of the system, considering the consent of its participants without causing policy conflict. For example, patients might enter personal information into a system to know whether or not they can get a refund for a surgery. The system provides only a refund for a specific type of surgery. When the system's policy rules change to incorporate other types of surgeries, the system should automatically update the rules and incorporate the changes without affecting the system's flow of information and processes.

### H. Policy of Data Sharing and Consent

In a healthcare system, the change in data collection and the usage practices require the respective policy to be revised and updated to reflect that change. In such a case, the system participant's consent is taken to be able to gather and process the collected data. Hence, a policy should explain the implications of granting or withholding consent to the participants so that data could be gathered and shared safely between authorized participants across and within systems. Thus, the data sharing and consent policy should consider:

*1) Comply With Standards:* Data sharing of healthcare participants should comply with legal requirements and security standards (e.g., ISO 27001, ISO/TC 215, ISO/TS 1360, ISO/IEC 29100). Data sharing might require knowledge of data formats, methods for securing data from malicious attacks, knowledge of available data management tools and software, and a complete understanding of the used applicable regulations and consent models. Thus, formal and comprehensive information security policies with respective implementation guidelines should be adopted to cover information security, access to information and systems, application security, information classification, and related security standards. The data sharing policy should provide guidelines, including constraints of regulatory requirements on when specific access conditions should be implemented.

*2) Policy Conflict Management:* Policy conflicts could arise due to conflicting requirements related to policy changes. Here, participants have to understand the updated policy to understand the changes, which is a complicated task. Thus, most system participants cannot make informed decisions about their privacy. Hence, analysis techniques (e.g., network analysis) [60][61] are used to explore the structural features of data sharing policy, map their relationships, and minimize the delay cost of sharing data within a system. However, network construction in healthcare systems might be complex and time-consuming due to the time required to analyze various activities, processes, sub-processes, and resource constraints within the system. The main challenge relies on the variety of policy representation forms at different levels of the hierarchy converted later to different forms for processing, which add complexity to the system structure and lead to inefficiency concerns.

One way to handle the conflict is by utilizing static conflict detection methods [62] to explicitly detect conflicting rules, compare all the paired rules, and analyze the conflicting probability of each pair of policy rules. However, this way is not affected as with the increasing number of rules in healthcare; the detection becomes inaccurate because of the increasing size of the policy rules set.

Another way is to classify policy rules into a classification tree [63] to detect the conflict by checking the conflict between nodes and their inherited rules. Nevertheless, this way consumes more memory, time, and space. It is less appropriate for a system that needs to predict continuous change. Moreover, the reproducibility of a tree is exceptionally sensitive as a slight change in policy rules could bring an enormous change in the tree structure.

A quantitative method using a linear combination of policy change is used [64][65]. The method further abstracted policy into a simple one and used a correlation matrix to detect the conflict according to the matrix. However, this method is not suitable for the dynamic nature of healthcare policy change as the size of the matrix depends on the number of rules; hence it is not suitable for a large number of rules.

Machine learning techniques (e.g., gossip learning technique, federated Learning algorithms, deep neural network) addressed the dynamic policy change challenge [66]–[68]. Here, performance and convergence scale degradation might happen because of the frequent change in policy and a lack of communication in healthcare, which results in an overfitting model. Moreover, such techniques under specific communication topologies could substantially impact the model's convergence speed in a decentralized environment when the speed is correlated with the data distribution. The training and execution of these models require extensive computational resources [61] which could not be used in limited environments with minimal performance.

Thus, there is a need to have standardization for data formats, existing protocols, and algorithms to enhance the reliability, interoperability, and modularity of the healthcare system and its components.

*3) Data Sharing Cost:* Sharing data accelerates health service delivery and promotes data reuse between healthcare participants. Here, healthcare participants' have preferences to share data within and across the systems in a distributed environment, which leads to an overload at the workflow and system levels. Such overload might cause a delay in data sharing among the system's participants and might increase the cost of sharing, including the time spent on a specific process to share the data (e.g., reviewing and analyzing data).

Furthermore, the cost of data sharing might be impacted by the dynamic change in the system's processes and information flow. Thus, the cost of sharing should be included in the policy. The cost should consider the time spent on an activity specific to sharing and reviewing data. The data sharing policy should enclose other costs such as hardware, software, data storage, and staff expenses, including a description of data charges, if any, and how these are calculated. The policy should address the requirements for metadata standard (e.g., HL7 Functional Model for EHRs) that varies by a health organization, considering the cost of storage, data access, and data management plan.

*4) Automated Policy Compliance:* An automated method could be applied to detect data change policy based on identifying and extracting the policy changes by considering precedence, relationships, and dependencies between policies to resolve the conflicts. The method could provide a self-adaptive mechanism that automatically adapts the updated policy to new data descriptions, rules, and regulations. The adaptive mechanism ensures that participants' rights and preferences are protected as their data are being shared in a distributed environment. Such a method could be promoted through pilot initiatives and ad-hoc regulatory guidance, which enable case-by-case deliberations throughout the diverse data types, and the various uses of data. Moreover, the changes in data sharing policy could be classified further to describe the severity levels that reflect the impact of changes on participants' privacy, policy consent, and data quality. The adopted method could be scalable to deal with the increasing data capacity and provide privacy-preserving data sharing across and within systems. Moreover, it could support various versions of data, provide different levels of data access depending on the version, and determine the best method of data sharing policy to ensure that system participants could use the data and prevent confusion, misuse, and misinterpretation.

## VI. Conclusions

Real-world digital healthcare systems are complex, hierarchical, and decentralized. In such systems, communications, dependency, precedence, and information shared between system clusters and their components, might be impacted by changes in the policies that govern allowed data flows within the system. Here, many barriers might strangulate the information flow and drastically impact the overall system's performance. Managing information flow and policy change in such an environment generates massive amounts of data that need to be aggregated, processed, and stored. At the same time, optimizing the communication cost when processing the data should be considered while minimizing the overhead in terms of CPU and Memory. The paper provided a technical view of the challenges, opportunities, and future research direction of the information flow and policy change/update in healthcare systems.

## References


[1] M. J. Page *et al.*, "The PRISMA 2020 statement: an updated guideline for reporting systematic reviews," *International Journal of Surgery*, vol. 88, no. April 2021, pp. 105906, 2021.

[2] F. Chaib and P. Garwood, "WHO releases first guideline on digital health interventions, 2019. [Online]. Available: https://www.who.int/news/item/17-04-2019-who-releases-first-guideline-on-digital-health-interventions. [Accessed: June 2022].

[3] Humanitas Group, "The timeline of digital health - Hunimed, 2022. [Online]. Available: https://www.hunimed.eu/news/the-timeline-of-digital-health/. [Accessed: June 2022].

[4] S. Kraus, F. Schiavone, A. Pluzhnikova, and A. C. Invernizzi, " Digital transformation in healthcare: Analyzing the current state-of-research," *Journal of Business Research*, vol. 123, no. February 2021, pp. 557–567, 2021.

[5] R. Sharma, and N. Kshetri, "Digital healthcare: Historical development, applications, and future research directions," *International Journal of Information Management*, vol. 53, no. August 2020, pp. 102–105, 2020.

[6] M. Senbekov *et al.*, "The recent progress and applications of digital technologies in healthcare: a review," *International Journal of Telemedicine and Applications*, vol. 2020, no. October 2020, pp. 1–18, 2020.

[7] Y. C. Lu, Y. Xiao, A. Sears, and J. A. Jacko, "A review and a framework of handheld computer adoption in healthcare," *International Journal of Medical Informatics*, vol. 74, no. 5, pp. 409–422, 2005.

[8] J. A. Powell, M. Darvell, and J. A. M. Gray, "The doctor, the patient and the world-wide web: how the internet is changing healthcare," *Journal of the Royal Society of Medicine*, vol. 96, no. 2, pp. 74–76, 2003.

[9] B. Müller, "Health and Care Futures Technology futures ," *Securing the future*, vol. 1, no. 1, pp. 1–26, 1999.

[10] H. S. Kim, I. H. Kwon, and W. C. Cha, "Future and Development Direction of Digital Healthcare," *Healthcare Informatics Research*, vol. 27, no. 2, pp. 95–101, 2021.

[11] World Health Organization, "Global strategy on digital health 2020-2025, 2021. [Online]. Available: https://apps.who.int/iris/handle/10665/344249. [Accessed: June 2022].

[12] World Health Organization and Regional Office for the Eastern Mediterranean, *Implementation guide for health systems recovery in emergencies: transforming challenges into opportunities.* IGO Publishing, 2020.

[13] C. Marcos, A. González-Ferrer, M. Peleg, and C. Cavero, "Solving the interoperability challenge of a distributed complex patient guidance system: a data integrator based on HL7's Virtual Medical Record standard," *Journal of the American Medical Informatics Association: JAMIA*, vol. 22, no. 3, pp. 587–599, 2020.

[14] M. B. Leavy and OMI, "Multinational Registries : Challenges and Opportunities," *Addendum to Registries for Evaluating Patient Outcomes: A User's Guide. 3rd ed. Rockville, MD: Agency for Healthcare Research and Quality*, vol. 3, no. February 2018, pp. 1–19, 2018.

[15] WHO and WORLDBANKGROUP, *Delivering Quality Health Services: A Global Imperative.* OECD Publishing, 2018.

[16] R. WHA74, "Strengthening WHO preparedness for and response to health emergencies," *Seventy-fourth World Health Assembly*, vol. 31, no. May 2021, pp. 1–14, 2021.

[17] Health IT, "What is an electronic health record (EHR)?, 2019. [Online]. Available: https://www.healthit.gov/faq/what-electronic-health-record-ehr. [Accessed: June 2022].

[18] C. Kirchberger, "Overview of the national laws on electronic health records in the EU Member States and their interaction with the provision of crossborder eHealth," *arXiv preprint arXiv:2108.04087*, vol. 22, no. July 2014, pp. 1–65, 2015.

[19] A. Kneck, M. Flink, O. Frykholm, M. Kirsebom, and M. Ekstedt, "The information flow in a healthcare organisation with integrated units," *International Journal of Integrated Car*, vol. 19, no. 3, pp. 1–10, 2019.

[20] M. J. A. Salomi and P. B. Claro, "Adopting Healthcare Information Exchange among Organizations, Regions, and Hospital Systems toward Quality, Sustainability, and Effectiveness," *Technol Invest*, vol. 11, no. August 2020, pp. 58–97, 2020.

[21] M. Hicks, S. Tse, B. Hicks, and S. Zdancewic, "Dynamic Updating of Information-Flow Policies," *Proceedings of the International Workshop on Foundations of Computer Security (FCS)*, vol. 20, no. June 2005, pp. 7–18, 2005.

[22] E. Daga, A. Gangemi, and E. Motta, "Reasoning With Data Flows and Policy Propagation Rules," *Semantic Web*, vol. 9, no. 2, pp. 163–183, 2018.

[23] G. L. Guernic and T. Jensen, "Monitoring Informa-

tion Flow," in *Workshop on Foundations of Computer Security-FCS'05*, 2006, no. May 2006, pp. 19–30.

[24] K. Chen, X. Guo, Q. Deng, and Y. Jin, "Dynamic Information Flow Tracking: Taxonomy, Challenges, and Opportunities," *Micromachines*, vol. 12, no. 8, pp. 1–16, 2021.

[25] M. Aydar, "Developing a Semantic Framework for Healthcare Information Interoperability," Ph.D. dissertation, Dept. Computer Science, Kent State Univ., Kent., 2015.

[26] S. H. Han, A. Nasridinov, and K. H. Ryu, "Information Flow Monitoring System," *IEEE Access*, vol. 6, no. May 2018, pp. 23820–23827, 2018.

[27] G. Michael, "Trust, Authority, and Information Flow in Secure Distributed Systems," Ph.D. dissertation, Faculty of the Graduate School, Cornell Univ., NY., 2020.

[28] B. Juliana, T. Webber, B. Euan, and V. Andreas, "A blockchain-based healthcare platform for secure personalised data sharing," *Public Health and Informatics: Proceedings of MIE*, vol. 31, no. May 2021, pp. 208–212, 2021.

[29] L. Tian, C. Pin, and S. K. Ting, "Blockchain-Based Healthcare Information Preservation Using Extended Chaotic Maps for HIPAA Privacy/Security Regulations," *Applied Sciences*, vol. 11, no. 22, pp. 1–16, 2021.

[30] M. K. Elghoul, "A Review of Leveraging Blockchain based Framework Landscape in Healthcare Systems," *International Journal of Intelligent Computing and Information Sciences*, vol. 21, no. 3, pp. 71–83, 2021.

[31] F. T. Lee, P. I. Chang, and S. T. Kung, "Blockchain-Based Healthcare Information Preservation Using Extended Chaotic Maps for HIPAA Privacy/Security Regulations," *Applied Sciences*, vol. 11, no. 22, pp. 10576, 2021.

[32] A. C. Myers and B. Liskov, "Protecting Privacy Using the Decentralized Label Model," *ACM Transactions on Software Engineering and Methodology, TOSEM*, vol. 9, no. 4, pp. 410–442, 2000.

[33] S. Mondal, and m. Nandini, "An efficient reachability query based pruning algorithm in e-health scenario," *Journal of Biomedical Informatics*, vol. 94, no. 1, pp. 103171, 2019.

[34] G. T. Nguefack, "Using bayesian networks to model hierarchical relationships in epidemiological studies," *Epidemiology and health*, vol. 33, no. June 2011, pp. 1–8, 2011.

[35] J. Rodon, and S. Leiser, "Exploring the formation of a healthcare information infrastructure: hierarchy or meshwork?," *Journal of the Association for Information Systems*, vol. 16, no. 5, pp. 1–43, 2015.

[36] C. Diana, and M. A. Martins, "Inconsistencies in health care knowledge," *IEEE 16th International Conference on e-Health Networking, Applications and Services Healthcom*, vol. 16, no. October 2014, pp. 37–42, 2014.

[37] R. Cruz-Correia *et al.*, "Automatic detection of patient data inconsistencies on integrated Health Information System," Ph.D. dissertation, Faculty of Medicine, Porto Univ., Porto., 2006.

[38] N. Angula and N. Dlodlo, "Towards a framework to enable semantic interoperability of data in heterogeneous health information systems in Namibian public hospitals," *Advances in Intelligent Systems and Computing*, vol. 721, no. January 2018, pp. 835–845, 2018.

[39] Health IT, "Health Interoperability 2030: Individual and Care Delivery Experiences, 2021. [Online]. Available: https://www.healthit.gov/topic/interoperability/health-interoperability-outcomes-2030. [Accessed: June 2022].

[40] Y. Wu, G. X. Yuan, and K. L. Ma, "Visualizing Flow of Uncertainty Through Analytical Processes," *IEEE Transactions on Visualization and Computer Graphics*, vol. 18, no. 12, pp. 2526–2535, 2012.

[41] Q. V. Dang, "Studying Machine Learning Techniques for Intrusion Detection System," in *International Conference on Future Data and Security Engineering*, 2019, pp. 411–426.

[42] Y. C. Chen, Y. J. Li, A. Tseng, and T. Lin, "Deep Learning for Malicious Flow Detection," *IEEE International Symposium on Personal, Indoor and Mobile Radio Communications, PIMRC*, vol. 2017, no. October 2017, pp. 1–7, 2018.

[43] M. Tiwari *et al.*, "Complete information flow tracking from the gates up," *Proceedings of the 14th International Conference on Architectural Support for Programming Languages and Operating Systems*, vol. 37, no. 1, pp. 109–120, 2009.

[44] M. Shana *et al.*, "Dynamic Information Flow Tracking for Detection of Advanced Persistent Threats: A Stochastic Game Approach," *ArXiv*, vol. abs/2006.12327, no. June 2020, pp. 1–15, 2020.

[45] P. Ammann, S. Jajodia, and P. Liu, "Recovery From Malicious Transactions," *IEEE Transactions on Knowledge and Data Engineering*, vol. 14, no. 5, pp. 1167–1185, 2002.

[46] J. Eder and W. Liebhart, "Workflow Recovery," *Proceedings - 1st IFCIS International Conference on Cooperative Information Systems, CoopIS*, vol. 19, no. June 1996, pp. 124–134, 1996.

[47] E. Aydemir and Y. Sahin, "Evaluation of Healthcare Service Quality Factors Using Grey Relational Analysis in a Dialysis Center," *Grey Systems: Theory and Application*, vol. 9, no. 4, pp. 432–448, 2019.

[48] Agency for Healthcare Research and Quality, "The CAHPS Ambulatory Care Improvement Guide: Practical Strategies for Improving Patient Experience, 2020. [Online]. Available: https://www.ahrq.gov/cahps/quality-improvement/improvement-guide/improvement-guide.html. [Accessed: June 2022].

[49] A. Niñerola, M. V. María-Victoria, and H. L. Ana-Beatriz, "Quality improvement in healthcare: Six Sigma systematic review," *Health Policy*, vol. 124, no. 4, pp. 438–445, 2020.

[50] L. Liqing, C. Zheng, and G. Xijuan, "Socially aware dy-

namic computation offloading scheme for fog computing system with energy harvesting devices," *IEEE Internet Things*, vol. 5, no. 3, pp. 1869–1879, 2018.

[51] S. Souravlas, and K. Stefanos, "Scheduling fair resource allocation policies for cloud computing through flow control," *Electronics*, vol. 8, no. 11, pp. 1348, 2019.

[52] Vigyanix, "Mirth Connect Series: Introduction, 2010. [Online]. Available: https://vigyanix.com/blog/mirth-connect-series-introduction/. [Accessed: June 2022].

[53] S. Jiang *et al.*, "Blochie: A Blockchain-based Platform for Healthcare Information Exchange," in *IEEE International Conference on Smart Computing, SMARTCOMP*, no. May 2019. IEEE, 2019, pp. 49–56.

[54] P. Pandey and R. Litoriya, "Implementing Healthcare Services on a Large Scale: Challenges and Remedies Based on Blockchain Technology," *Health Policy and Technology*, vol. 9, no. 1, pp. 69–78, 2020.

[55] S. Tanwar, K. Parekh, and R. Evans, "Blockchain-Based Electronic Healthcare Record System for Healthcare 4.0 Applications," *Journal of Information Security and Applications*, vol. 50, no. February 2020, p. 102407, 2020.

[56] K. N. Griggs *et al.*, "Healthcare Blockchain System Using Smart Contracts for Secure Automated Remote Patient Monitoring," *Journal of Medical Systems*, vol. 42, no. 7, pp. 1–7, 2018.

[57] H. A. El Zouka and M. M. Hosni, "Secure IoT Communications for Smart Healthcare Monitoring System," *Internet of Things*, vol. 13, no. February 2021, pp. 1–14, 2021.

[58] J. Kharel, H. T. Reda, and S. Y. Shin, "An Architecture for Smart Health Monitoring System Based on Fog Computing," *Journal of Communications*, vol. 12, no. 4, pp. 228–233, 2017.

[59] Deloitte, "2022 Global Health Care Outlook Are we finally seeing the long-promised transformation?," *Deloitte Global Health*, vol. 4, no. 2022, pp. 1–52, 2022.

[60] R. Khelf, and N. Ghoualmi, "Intra and inter policy Conflicts Dynamic Detection Algorithm," *Seminar on Detection Systems Architectures and Technologies DAT*, vol. 1, no. February 2017, pp. 1–6, 2017.

[61] O. Salem, Y. Liu, M. Ahmed, and B. Raouf, "Online anomaly detection in wireless body area networks for reliable healthcare monitor," *IEEE journal of biomedical and health informatics*, vol. 18, no. 5, pp. 1541–1551, 2014.

[62] E. Syukur, S. W. Loke, and P. Stanski, "Methods for policy conflict detection and resolution in pervasive computing environments," *Policy Management for Web workshop in conjunction with WWW2005 Conference*, vol. 14, no. May 2005, pp. 13–20, 2005.

[63] L. Xueting, "A Method of Conflict Detection for Security Policy Based on B+ T," *IEEE Fourth International Conference on Data Science in Cyberspace DSC*, vol. 4, no. June 2019, pp. 466–472, 2019.

[64] X. Yuping, and W. Shengli, and Y. Zhan, "Statistical analysis of the linear combination method," *Journal of Computational Information Systems*, vol. 11, no. 18, pp. 6615–6620, 2015.

[65] J. Malczewski, "On the use of weighted linear combination method in GIS: common and best practice approaches," *Transactions in GIS*, vol. 4, no.1, pp. 5–22, 2000.

[66] L. Giaretta, "Pushing the Limits of Gossip-Based Decentralized Machine Learning," Master thesis., School of Electrical Engineering and Computer Science, KTH., Sweden., 2019

[67] R. Miotto, F. Wang, S. Wang, X. Jiang, and J. T. Dudley, "Deep learning for healthcare: review, opportunities and challenges," *Briefings in bioinformatics*, vol. 19, no. 6, pp. 1236–1246, 2018.

[68] A. A. Abdellatif *et al.*, "Reinforcement Learning for Intelligent Healthcare Systems: A Comprehensive Survey," *arXiv preprint arXiv:2108.04087*, vol. 1, no. August 2021, pp. 1–31, 2021.

# Conceptual Model of the Application of the ABA Method in Alzheimer's Treatment Supported by Data Science

Priscila Firmiano Parede
Religare Innovation Lab
Marília, Brasil
e-mail: priscila.parede@conectamente.net.br

Elvis Fusco
Fundação Shunji Nishimura de Tecnologia
Pompeia, Brasil
e-mail: elvisfusco@gmail.com

Caio Saraiva Coneglian
Universidade de Marília UNIMAR
Marília, Brasil
e-mail: caio.coneglian@gmail.com

***Abstract* - Alzheimer's Disease (AD) is characterized as a neurodegenerative disorder that causes gradually progressive cognitive and functional deficits and behavioral changes. Checking for symptoms of cognitive loss and memory loss, behavioral symptoms, functional decline, and cognitive testing remain the cornerstone of clinical diagnosis and treatment of AD patients. In view of the questions raised from the experience of the researchers, as well as the results achieved by the Applied Behavior Analysis (ABA) method in patients with Autism Spectrum Disorder (ASD), and adopting an innovative perspective of the ABA method aimed at the treatment of patients with AD, we envision the possibility of supporting the ABA method with the use of Data Science (DS) techniques to enable analysis and learning based on the information generated in this process, in search of better quality of life and cognitive, speech-language and psychomotor advances in patients with AD, featuring an important interdisciplinary approach.**

***Keywords - Alzheimer's Disease; Applied Behavior Analysis; Data Science; Conceptual Model.***

## I. INTRODUCTION

Alzheimer's Disease is characterized as a neurodegenerative disorder that causes gradually progressive cognitive and functional deficits and behavioral changes [7]. The specialized literature indicates that its most common cognitive symptoms include deficits in short-term memory, executive, visual and spatial dysfunction, difficulty in carrying out daily activities, even total disconnection from reality [9]. Current studies are converging in stating that the verification of symptoms of cognitive loss, the identification of behavioral symptoms and functional decline and the performance of cognitive tests remain as the basis of the clinical diagnosis and treatment of patients with AD [11]. Individuals with Autism Spectrum Disorder (ASD) or with AD have specific deficits in their ability to learn procedural skills that are explained by the loss of their motor coordination. Thus, we can say that an interdisciplinary team that works with the Applied Behavior Analysis (ABA) method will be able to offer specific stimuli to patients with AD, as is already the case with individuals with ASD, whether they are professionals in Psychology, Speech Therapy, Psychomotricity, Physiotherapy, among others. The use of the proposed tests becomes fundamental in the evaluation of the patient with AD and in the validation of the impact of the ABA treatment, using parameters of comparison of the affected cognitive and motor areas. It is not a question of comparing the use of the ABA method with other methods, but of verifying its feasibility and its application in patients with AD, which has not yet been reported in the literature.

Conjugating with the results of the studies of Castillo et al. [7] and Perakslis et al. [11], we believe that there is no universal treatment for AD, but it is possible to adopt an interdisciplinary and preventive intervention, consisting of psychosocial treatments, behavioral therapy and drug treatment. In this sense, we propose the application of the ABA method in the treatment of Alzheimer's Disease, and based on recent studies that indicate a constant growth in research investment that aligns technologies in the area of Data Science applied to the health area, we envision the possibility of aligning the ABA method to data analysis to support and validate the treatment of AD patients.

The general objective of this study is to analyze the potential of the ABA method in the treatment of Alzheimer's Disease with the support of an informational platform based on Data Science (DS) techniques. Specifically, we intend to: a) study the feasibility and effectiveness of the ABA method in patients diagnosed with Alzheimer's Disease, with a view to expanding quality of life and promoting cognitive, speech-language and psychomotor advances and b) developing a prototype composed of a digital platform using data analysis techniques and algorithms linked to the application of the ABA method to support the treatment of AD patients as a proof of concept of the proposal's feasibility.

This article has the following structure: Section 2 presents a review of the literature on the concepts relevant to the research, with emphasis on the Applied Behavior Analysis method and on Alzheimer's Disease. Section 3 addresses the methodological procedures adopted in the study and the results achieved. Finally, the conclusion is presented.

## II. LITERATURE REVISION

Behavior Analysis is a science that is dedicated to the study of the elements that affect behaviors. The use of the principles of this science to solve socially relevant demands is called Applied Behavior Analysis – ABA [1]. The ABA method can be applied in different contexts such as ASD treatment, education, economics, psychological clinic, sports performance, among others.

Despite being used in several areas, there was a significant growth in the use of ABA in the area of ASD, especially Lovaas [2], who carried out the first study of ABA applied to the treatment of this type of disorder. The method consists of individual and intensive teaching of skills necessary for the individual to acquire independence and better quality of life. Among the skills taught, those that focus on behaviors that interfere in the development and integration of the individual diagnosed with ASD stand out: a) social behaviors, such as eye contact and functional communication; b) academic behaviors, as prerequisites for reading, writing and mathematics; c) activities of daily living, such as personal hygiene; d) reduction of negative behaviors, such as aggression, stereotypies, self-injury, verbal aggression and escapes.

The literature indicates improvement in the development of participants after undergoing the ABA method, especially when faced with other types of non-intensive therapies or not anchored in Behavior Analysis, especially when the intervention was performed early [3][4][5][ 6]. Based on these studies, which reinforce the success of the ABA method in the treatment of patients with ASD and, as in the studies carried out with patients with AD [7][8], we adopted part of the methodology presented in this study, namely, the of tests in the areas of Speech Therapy, Psychology, Psychomotricity and Physiotherapy.

Castillo et al. [7] exposed a group of AD patients to psychosocial treatment, aimed at individuals with mild and moderate dementia. The treatment included physical activities, sound identifications, childhood explorations, names, foods, current discussions, identification of faces/scenes, word associations, increased creativity, categorization of objects, development of senses, among others. The results obtained were "improvement in orientation, a better understanding and adaptation of their environment, potentially leading to an increase in the feeling and perception of control, self-efficacy and self-esteem", directly impacting the quality of life of patients and their families [7][9].

Giving an innovative perspective to the use of the ABA method in the treatment of AD, we propose to reconcile it with techniques from the DS area to support the treatment of patients through data analysis and the learning generated based on the information obtained in this process, in search of better quality of life and cognitive, speech-language and psychomotor advances, thus justifying other methodological aspects of the study.

Recent literature points out that medical treatments bring together cutting-edge technology, such as systems for remote surgery, exoskeletons or wearables for the elderly to be monitored remotely, for example. IoT, Big Data, AI, nanotechnology and robotics, for example, are available to health professionals for hospital care, diagnosis, staging and monitoring of treatments. The reason for applying DS in medicine is to obtain improved services through the use of data and information analysis. These services are classified in two aspects: the first is the diagnosis and prognosis of the disease, and the second is its treatment and definitive cure [10].

The volume of relevant data on AD generated by the tests that will be applied will generate the need for DS techniques to deal with the informational complexity of these environments. Data Science can provide insight and pattern detection and/or insights for possible decision making. Constant progress in the development of predictive modeling and analysis, digital sensors, advanced monitoring techniques and their analytical capabilities are prepared to allow the real-time study of patients and caregivers in a continuous way in all stages of the disease [11]. A new perspective is needed to manage and process this information, while organizing these resources into new results.

In another study, Krumholz [12] highlights how Data Science techniques and data analysis in Big Data scenarios can contribute to the generation of new knowledge essential for predictions and improve the performance of health treatments. Furthermore, the author points out that the use of Data Science techniques can significantly improve the power of observation for treatments in the health area, precisely because of the amount of data obtained and the use of these data to obtain new knowledge and define patterns that are not found by humans.

## III. METHODOLOGY

Considering that this is a feasibility study, we started with the Systematic Literature Review (SLR) in search of information about the scope of the study, in order to synthesize the production of knowledge about Alzheimer's Disease, the application of the ABA method, the use of Data Science in the health area and, in particular, in the treatment of AD. By retrieving the results of the research already carried out on these subjects, we find:

- Case studies and applications of the ABA method, to establish possible relationships between its application in patients with ASD and patients diagnosed with AD in terms of cognitive gains;
- Case studies and applications of tests and scales in patients with AD in the areas of Psychology, Speech Therapy, Physiotherapy and Psychomotricity, to establish possible relationships between their application, cognitive gains and improvement in quality of life;
- Case studies, techniques, applications and DS models applied in learning and improving the treatment of AD based on massive environments of data generated by the treatment and the tests mentioned in the previous activities;

- Case studies and identification of methodologies and technologies for modeling, merging, mining, management and data analysis, applied to the representation and processing in digital informational environments of data analysis systems, based on information from the capture of AD treatment with the application of the ABA method.

The SLR followed the precepts described by Higgins and Green [13] and included works published in the period from 2000 to 2021, with regard to works on ABA and DA and in the period from 2010 to 2021, with regard to works on SD and national and international journals were searched.

We used the Multiple Case Study (MCE) research method, as proposed by Yin [14]. The method is characterized by simultaneously studying more than one particular case representing a set of comparable or equivalent cases. It is an "empirical investigation that investigates a contemporary phenomenon within its real-life context, especially regarding the boundaries between the phenomenon and the context that are not clearly defined", seeking to know the how and why of a specific situation [ 14].

The MCE is composed of patients from the Religare Clinic – Rehab Center. The Religare Clinic is located in the city of Santo André, metropolitan region of the State of São Paulo, Brazil. Diagnosed with mild or moderate AD, identified and chosen intentionally, for Alzheimer's Disease submitted to the ABA method, patients will be submitted to the application of tests and scales in the areas of Physiotherapy, Psychomotricity, Psychology and Speech Therapy, to assess their cognitive performance and improvements in quality of life and to validate the ABA method in the treatment of these patients. The tests will be applied in a future stage of the research. In MCE, data analysis can be performed using various techniques and tools, such as the use of computer programs, tabulations, categorization, tests and combination of evidence and data collected in SLR. At this stage of the research, we will make use of the following strategies: analysis in the light of SLR; analysis based on the results achieved with the tests and scales applied to patients; analysis based on the information provided by the metadata standard implemented to represent AD treatment data. The use of disruptive computing technologies using Data Science resources will expand the possibilities of treatment through the digitization and automation of processes that increase the ability to exponentiate the analysis of results.

It is proposed to use massive data analysis techniques generated from the diagnosis history, application of the ABA method and tests in related areas to support treatment evaluation plans and improvements in the adequacy of the ABA method in the treatment of AD.

Figure 1 presents the conceptual model of the proposal that results in the development of the prototype that will be validated through a proof of concept of the presented methodology.

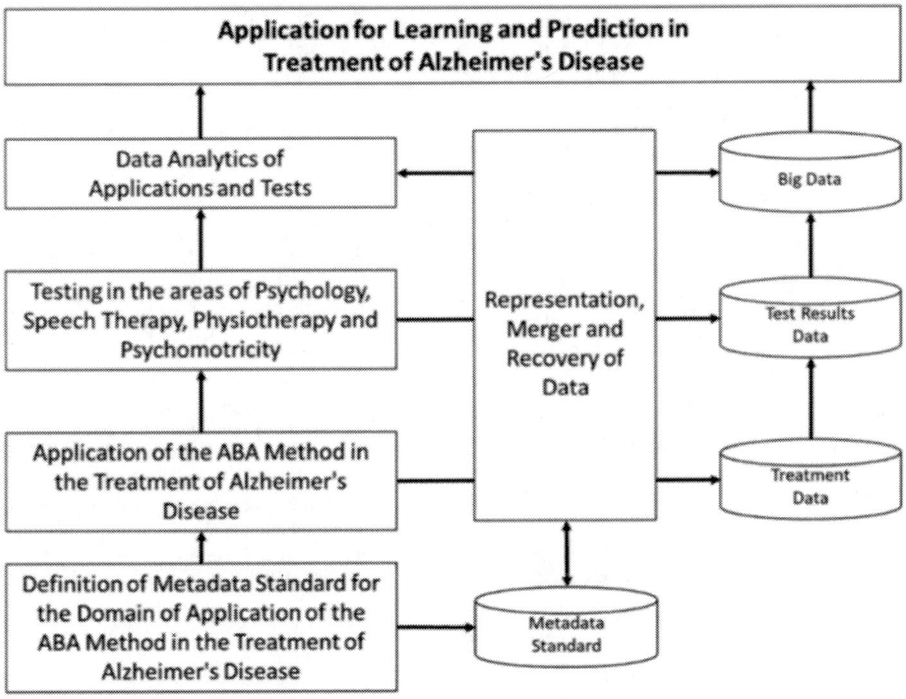


Figure 1: Conceptual Model of Application of the ABA Method in Alzheimer's Treatment Supported by Data Science

The main purpose of the conceptual model is to capture information requirements from a research point of view. As an important instrument used to represent the prototype to be developed, the construction of a Conceptual Model should not be limited only to the need to represent the prototype, but rather to develop global views of the entire research process. This can be seen in Table I that specifies the components that represents the Conceptual Model.

TABLE I. CONCEPTUAL MODEL COMPONENTS

| Category | Component | Especification |
|---|---|---|
| Data Layer | Metadata Standard | Standardized data scheme of informational elements describing the characteristics of AD and treatment methods. |
| Data Layer | Treatment Data | Data extracted from the treatment of AD. |
| Data Layer | Test Result Data | Results obtained through tests applied after treatment |
| Data Layer | Big Data | Database formed by several information sources obtained from diagnosis, treatment and tests. |
| Process Layer | Definition of Metadata Standard | Specification of the conceptual data schema of the proposed model. |
| Process Layer | Application | Treatment of patients with AD using the ABA Method. |
| Process Layer | Testing | Application of tests in patients treated with the ABA Method. |

To support the digital informational application to be developed to validate this proposal, a metadata standard is defined that simplifies and standardizes the data that will be generated in the application phases of the ABA method in the treatment of AD and in the application of the tests. From this metadata standard, a database with Big Data characteristics will be built from the generation of data in each subsequent phase, until the moment of analysis of the generated data and transformation into information for the application of learning and prediction that improve the AD treatment process. It is expected that with the implementation of this model, there will be an improvement in the treatment of AD through the application of the ABA method supported by indicators and knowledge generated by applications in the area of Data Science based on Big Data environments.

## IV. Conclusion

The partial results indicate that the technological structure applied in the AD treatment process will allow innovation in the treatment methodology through the ABA method applied in AD and innovation in the use of disruptive computational technologies in the process of evaluation, testing and learning of AD treatment. The feasibility study of the techniques and activities of the ABA method to be adopted in the treatment of patients with AD is under development and a protocol for the application of tests and scales in the areas of Psychology, Speech Therapy, Psychomotricity and Physiotherapy able to validate the ABA method in the treatment of patients with AD.

In the end, it is expected to develop a prototype of DS application based on the information generated by the tests for learning and predicting the treatment of AD with ABA method resources, to subsidize the proof of concept that supports this proposal.

## References


[1] J. Todorov and E. Hanna. "Behavior analysis in Brazil". Psicologia: Teoria e Pesquisa, Brasília, v.26, n. 25 anos, 2010, pp.143-154.

[2] O. Lovaas. "Behavioral treatment and normal educational and intellectual functioning in young autistic children". Journal of Consulting and Clinical Psychology, v.55, n.1, 1987, pp. 03-09.

[3] S. Eldevik, et al. "Meta-analysis of early intensive behavioral intervention for children with autism". Journal of Clinical Child & Adolescent Psychology, v.38, n.3, 2009, pp. 439-450.

[4] Z. Warren, et al. "A systematic review of early intensive intervention for autism spectrum disorders". Pediatrics, v.127, n.5, 2011, pp. e1303-e1311.

[5] C. Shepley and d J. Grisham-BrowN. "Applied Behavior Analysis in Early Childhood Education: An Overview of Policies, Research, Blended Practices, and the Curriculum Framework". Behav. Analysis Practice. 12, 2019, pp. 235-246.

[6] D. S. Im "Treatment of Aggression in Adults with Autism Spectrum Disorder: A Review". Harv Rev Psychiatry. 2021 Jan-Feb; 29(1): pp. 35-80.

[7] C. Miranda-Castillo, F. M. Tapia, A. R. Herrera, F. M. Ghigliotto and L. S. Guerra. "Implementation of a cognitive stimulation program in people with Alzheimer-type dementia: a pilot study in elderly Chileans". Universidad Psicología, Bogotá, v. 12, n. 2, 2013, pp. 445-455.

[8] T. Kihara, et al. "Galantamine modulates nicotinic receptor and blocks abeta-enhanced glutamate toxicity". BiochemBiophys Res Commun. 325(3), 2004, pp. 976-82.

[9] L. D. Apostolova. "Alzheimer Disease". Continuum (Minneap Minn), 22(2), 2016, pp. 419-434.

[10] R. Pandya; et al. "Buildout of Methodology for Meticulous Diagnosis of K-Complex in EEG for Aiding the Detection of Alzheimer's by Artificial Intelligence". Augment Hum Res. 5:3, 2020, pp. 01-09.

[11] E. Perakslis, H. Riordan, L. Friedhoff, A. Nabulsi and E. Merlo Pic. "A call for a global 'bigger' data approach to Alzheimer disease". Nat. Ver. Drug. Discov. 18, 2019, pp. 319-320.

[12] H. M. Krumholz. "Big data and new knowledge in medicine: the thinking, training, and tools needed for a learning health system". Health Affairs, v. 33, n. 7, 2014, pp. 1163-1170.

[13] J. P. T. Higgins and S. Green. "Cochrane handbook for systematic reviews of interventions", version 5.1.0. Oxford: The Cochrane Collaboration; 2011.

[14] R. K. Yin. "Case study: planning and methods". 2a ed. Porto Alegre: Bookman, 2001.

# A Practice-oriented Approach to Participatory Design

## Designing Health Information Systems for Healthcare Networks

Øivind Skeidsvoll Solvang [a], Sonja Cassidy [a], Ove K. Lintvedt [b], Conceição Granja [b,c] and Terje Solvoll [b,c]

[a]Department of Architecture and Innovation
Helse Vest IKT AS
Bergen, Norway

[b]Norwegian Centre for e-health Research
University Hospital of North Norway
Tromsø, Norway

[c]Faculty of Nursing and Health Sciences
Nord University
Bodø, Norway

*e-mail:* oivind.skeidsvoll.solvang@helse-vest-ikt.no[a], sonja.cassidy@helse-vest-ikt.no[a], ove.lintvedt@ehealthresearch.no[b], conceicao.granja@ehealthresearch.no[b,c], terje.solvoll@ehealthresearch.no[b,c]

***Abstract*—Active user involvement is crucial to designing a Health Information System (HIS) consistent with users' expectations and needs. Many design projects use Participatory Design (PD), involving users as active co-designers, to promote the democratization of work and empower the workers. While this design approach has worked in local contexts, there is an open question on how to broaden participatory design to healthcare networks in need for coordinating their services for patient-centric care. In patient-centric healthcare, various providers at different healthcare levels must coordinate their actions. In this paper, we discuss different challenges when PD applies to a healthcare network and how practice-oriented approaches can contribute to mitigating these challenges.**

***Keywords-Participatory Design; Practice Theory; Health information systems.***

### I. INTRODUCTION

The increasing demand for healthcare services in several countries [1] puts pressure on health care providers [2] to increase throughput and quality of care, without increasing resources [2]. This pressure has caused more workload, stress, and dissatisfaction among clinicians, resulting in burnout and higher turnover [3]. Health Information Systems (HIS) can have a role to play in supporting clinicians to reduce their workload. However, many information systems fail when introduced [4] due to user resistance as a central contributing factor [5]. For the context of this paper, a HIS is a general term for digital information systems that assist clinicians in collecting, processing, and communicating health information internally and externally [6].

Recent studies on barriers and facilitators to HIS adoption indicate a compatibility gap between clinicians' expectations and the final solution [7]. Such a gap might occur when the final solutions are not consistent with their values, needs, past experiences, and work practices [8]. However, clinicians' needs and knowledge are sometimes hidden or hard to articulate [9]. In other situations, clinicians are not aware of their needs, or the needs change during the design process [9]. Thus, to narrow the gap between their real needs and the final solution, it is crucial to involve clinicians actively during the design process [4][7]. In this paper, we follow the definition by Ehn [10] that design is "a concerned social and historical activity in which artifacts and their use are anticipated; an activity and form of knowledge that is both planned and creative, and that deals with the contradiction between tradition and transcendence".

Participatory Design (PD) has a particular attention to hidden knowledge, needs and advocates an active involvement of users as co-designers [11]. Through active participation, users build competencies, are empowered and stimulated to meaningful work [12][13]. PD involves design and redesign of workplaces, jobs and technologies [14]. PD is a mutual learning process where participants and designers bring their unique expertise [13]. The designers learn about the work practices from the practitioners to give advice of future technological possibilities. The practitioners learn from the designers to be empowered to design their future work.

Much focus has been on situated and local learning in PD, with limited considerations of evolving the learning to broader arenas [14]. The coordination of activities and the need to share knowledge among practitioners across providers and levels of care (healthcare network) increases as the attention on patient-centric care evolves. The question is how to approach PD in a healthcare network, with boundaries among providers and practitioners, potentially containing an uneven distribution of responsibilities, resources, power, and decision rights.

Practice-oriented approaches also study knowledge emerging through mutual interaction. However, practices are the basic unit of analysis in these approaches. New ways of understanding social and organizational phenomena are possible when studying the performance of practices' actions [15]. Through practice-oriented approaches, situated relations between practices and the arrangements keeping the practices together are analyzed and used for further studies of variations in practices at other sites and contexts [16]. In addition, by changing the arrangements that keep the practices together, this approach is also a resource for change [17].

Currently, we are working on a project, Valkyrie [18], where the aim is to promote healthcare service coordination for patient-centric care across healthcare levels. The ambition is to collect, coordinate, and present health data records from several heterogeneous Electronic Health Record (EHR) systems as a coordinated, standardized, virtual electronic health record system. The project involves clinicians from different autonomous practitioners across primary care (general practitioners and municipalities) and specialist care (hospitals). While boundaries exist between the providers, the broader healthcare network is dynamic and changes according to the individual patients' pathway. We expect to uncover conflicting needs among providers and practitioners during this project. Thus, active user involvement in the design process is crucial for the success of this project.

Many design research methods rely on user involvement. To our knowledge, PD is the only approach with the active participation of users co-designing their work future. To meet the challenges in the Valkyrie project, we decided to focus on the PD approach.

However, because of the known weaknesses of the method in PD, we wanted to see if a practice-oriented approach could help enrich the method to avoid some of its weaknesses. From earlier research, we were aware of the use of practice-oriented approaches in telemedicine [15], Sociomateriality [16], as well as in the study of educational and nursing practices [19]. However, we were unaware of studies using practice-oriented approaches in PD to expand the design process in a healthcare network, and to help overcome known weaknesses in PD.

In this paper, we discuss different challenges with using PD in a healthcare network and elaborate on how practice-oriented approaches can contribute to mitigating these challenges.

The rest of this paper is organized as follows. Section II describes methods and results from the brief literature search. Section III starts with a description of PD and practice-oriented approaches, and ends with a discussion of how practice-oriented approaches can help mitigate PD's challenges. Finally, Section IV gives a summary of the research and highlight its contribution and future work.

## II. Methods And Result

We conducted a search in Scopus, without any limit on year of publication, with the query string: TITLE-ABS-KEY ("participatory design" AND practice AND ( healthcare OR "health care") AND ("information system" OR "electronic health record system" OR telemedicine )), limited to title, abstract, and keywords. To verify our search, we used the same search terms in Google Scholar.

The search in Scopus identified 33 papers. We screened the papers based on their abstracts, and we decided not to include any of them since the abstracts did not indicate any use of practice-oriented approaches in PD.

The search in Google Scholar identified 19 papers. We screened the papers based on their abstracts, and we decided to include one paper. This paper introduces a practice-oriented approach to bridge ethnography with design research, including PD, for the healthcare domain [20].

Because of the limited results from the search, we did not consider any inclusion or exclusion criteria necessary.

## III. Discussion

The brief literature search conducted for this study did not show any significant use of practice-oriented approaches combined with PD. However, there are chances that we could have identified more papers by using other search terms. Searching for "participatory design" or "practice" would have resulted in more papers. However, we assume that such an approach would not have helped us identify more papers intersecting the two research fields.

This section discusses some challenges with using PD in a healthcare network and elaborates on how practice-oriented approaches can mitigate some of these challenges.

### *A. Participatory Design*

PD emerged as a political protest movement in Norway [12], advocating active user involvement in design as means to increased empowerment [12], build competencies and stimulate meaningful work [13]. This design approach views knowledge as being created through the interaction among people, practitioners and artifacts [12]. It is pragmatic in that the design should also contribute to improvements for the research subjects [12]. Thus, PD is action oriented. In order to design for change, envisioning and enacting possible futures are necessary. As PD has evolved, techniques and tools are developed, such as stakeholder analysis, to find relevant stakeholders influencing the design or influenced by the design [13]. Tools and techniques such as personas, scenarios, mock-ups, simulations, future workshops and cooperative prototyping are commonly used in PD to help participants substantiate future possibilities with practical discussions [21].

Three basic steps for mutual learning and design are usually present in PD. First is the initial exploration of work, where practitioners and designers meet and explore workflows, routines and how technologies can support practitioners in their practice. Second is the discovery process, envisioning the future workplace and prioritizing tasks. The last step is prototyping an envisioned solution, involving several iterations of sketches, drawings, and descriptions, gradually evolving the prototype in a lab or at the workplace. The iterations allow participants to examine, reflect and discuss incremental redesigns at each iteration [12].

While PD can positively affect design outcomes, it can also introduce some challenges. Democratization of work, for instance, entails that each participant's voice is equally valued, even those that offend others or argue out of self-interest [13]. There is also a risk that certain practitioners use power to influence design decisions, and such challenges can cause tension and conflicts with not necessarily an optimal design result.

Another challenge is to establish an environment for mutual learning. In PD, the activities that directly engage participants in the design are in the foreground, while understanding and learning becomes a natural part of the design work [22]. In contrast, ethnography starts with understanding work practices, which involves detailed workplace studies. As PD has evolved, designers have started to use techniques from ethnography to learn about current problems and work practices before they meet participants. However, it is not always clear when ethnography stops and design starts, and how designers can benefit from the contributions of ethnography [22].

The third challenge with PD is to involve the suitable candidates as participants and co-designers to ensure that the information system to be designed is well integrated into practitioners' work practices [21], thus more likely to be accepted as valuable.

The fourth challenge with PD is how to broaden the learning from one particular site to a design of HIS that is perceived valuable and accepted by practitioners in broader areas [21]. With the growth in providers and practitioners, variances in work practices and needs are likely to increase. The same applies to the risk of conflicts and tension among practitioners and providers. Finding participants that can look beyond their work are essential [21]. However, this also calls for attention to differences in organizational structures [23], such as roles, regulations, hierarchies and power relations, internally and across providers.

### B. *Practice Theory*

Practice theory is one way to study organizational phenomena. Practice theory views organizations as consisting of practices that produce an outcome. A healthcare organization, in this view, is both the physical place and the result of the practices' work activities [15]. Following this view, a hospital, for instance, consists of different practices. All perform necessary actions at the physical place for the outcome of a healthy patient, such as cleaning rooms and beds, diagnosing the patient, delivering food to the patient and so on. The actions are of particular interest in this research field since actions together constitute the practice [24].

Practice-oriented approaches do not follow a single, unified theory. However, they all have in common that practices, not individuals or practitioners, are the basic unit of analysis for understanding organizational phenomena [15]. New ways of understanding social and organizational phenomena are possible by studying the performance of actions by practices [15]. Another aspect that practice theories generally have in common is the recognition of materiality, communication, and symbols as essential parts of the practice composition [19].

Practice theories usually follow Schatzki's site ontology, stating that practices consist of language (sayings) and activities (doings), always situated within a site or sites [25]. Knowing is sustained and manifested in practice and through practices in the happening (situated) [26], which marks parallels to PD, with its focus on knowledge construction through the interaction among people and artifacts [12].

Sociomateriality is a practice-oriented notion that views the performance of activities that mutually shape and reshape materiality (physical and digital material) and social phenomena [23]. In this perspective, the boundaries between materiality and social phenomena are inseparable and not pre-defined [27]. Through sociomaterial practice, which is the space for human and material agencies, objects emerge as sociomaterial configurations [23]. Caused by differences in performance, the sociomaterial configurations can take different forms, with varying social and material consequences, intended or unintended [27]. Following the sociomaterial tradition, Nicolini studied the mutually shaping of practices and telemedicine [15]. Orlikowski found that organizational norms affect people's decisions on using information systems, but when in use, people start shaping the system and seek new possibilities [16][23].

While Sociomateriality often studies current relations, design is future-oriented [20]. By introducing Sociomaterial-Design, Bjørn and Østerlund are concerned with bringing Sociomateriality closer with practical design research, including PD, for the healthcare domain [20]. Their studies show how Sociomaterial-Design can bridge ethnography and design, which can be a valuable contribution to design research.

However, Sociomaterial-design and Sociomateriality, in general, suffer from treating organizational structures implicit. According to Leonardi [23], sociomaterial practice is similar to a technical subsystem from a socio-technical system perspective. Leonardi [23] reminds us that organizations are socio-technical systems consisting of a technical and social subsystem. The social subsystem consists of constructs such as roles, statuses, hierarchies, and power relations. While humans and materiality mutually shape each other in a technical subsystem, there is also an ongoing mutual shaping of the social and technical subsystems. In healthcare, the social subsystem is strong and affects knowledge sharing and collaboration between organizational and professional boundaries [28]. When designing a HIS for a health care network, it is crucial to understand both the social and the technical subsystems and how they affect each other.

| Actions | Intersubjective spaces | Practice architecture |
|---|---|---|
| Sayings | **Semantic space** (medium of **language**) | **Cultural-discursive arrangements** |
| Doings | **Physical space-time** (medium of **activity** and **work**) | **Material-economic arrangements** |
| Relatings | **Social space** (medium of **power** and **solidarity**) | **Social-political arrangements** |
| Actions performed by practices and bundled together in projects. | | Arrangements that enable and constrain interactions, bundled together in practice landscapes and traditions at particular sites. |

Figure 1. A simplified illustration of the theory of practice architectures [19]

In contrast to Sociomateriality, the theory of practice architecture expands on Schatzki's site ontology, making power and politics (Relatings) explicit, together with language (Sayings) and activities (Doings). Developed initially for studying educational practices, the theory of practice architecture has gradually evolved in other practice fields, such as nursing, teacher mentoring and professional learning at universities [19]. To our knowledge, no studies so far have used this theory in PD.

The theory of practice architecture (Figure 1) is domain-neutral and represents a view of what constitutes and shapes the practices. Practices in this theory, are human actions composed of what they say (Sayings), what they do (Doings) and how they relate (Relatings) [19][24]. These actions are interconnected and not reducible to one of these actions in isolation [29].

Projects bundle actions (left side in Figure 1) and contain the intentions that motivate the practice, the aims to achieve, and the practitioners' dispositions. When a nurse is engaged in nursing, one such project is to support the recovery of patients after surgery. The specific activity is justified through particular sayings and social relatings, with the intention and aims of the nursing practice. Through actions, aim, intention and disposition, practices can be examined and give insight into how and why others perform the practice differently.

There is also a continuous mutual shaping between the practices and the context. The context manifests itself in aspects of what is sayable, doable, relationships, rules, regulations and expectations. Through the theory of practice architecture, context emerges in the foreground. It becomes tangible through the practice architecture consisting of three interconnected arrangements, cultural-discursive, material-economic and social-political (right side in Figure 1) [19][24]. The practise architecture are the preconditions, traditions, constraints and enablement of the practices, by occurring in or brought to a particular site [19]. Transformation of practices' actions can also occur by bringing in new arrangements or creating new arrangements at the site [30]. For instance, rearranging the inside of a hospital building or creating a new information system can promote changes in the practices' actions.

The relations between practices and practice architecture emerge through three intersubjective spaces (Figure 1): interrelated semantic space, physical space-time, and social space. Through these spaces, practitioners come together in a shared language, material reality, power and solidarity [17] for mutual learning, acceptance and maintenance of the practice, and shaping and reshaping the practice [29].

The theory of practice architecture can be a valuable theoretical resource providing a concise language for interpretation and description [17]. It can explain situated dynamics and relations at one particular place, transferrable to study relations in other situations and contexts [16]. The theory can also serve as an analytical tool for identifying actual empirical connections between practices and the practice architecture, focusing attention on local variations in the mutual shaping of practices and the arrangements. Finally, it can be used as a resource for change, transforming the arrangements that keep the practices together [17].

### *C. How can practice-oriented approaches contribute to mitigate the challenges of participatory design in Valkyrie?*

#### *1) Equal voice and power relations*

Democratization of work entails that every participant's voice is equally valued. Researchers have criticized this core value for letting those voices offend others and argue out of self-interest to be equally valued [13]. Since practice-oriented approaches use practices as the basic unit of analysis, they can help participants shift perspectives and view organizational phenomena from other angles. This shift in focus may help practitioners broaden their view, thus reducing arguments out of self-interest. However, we should also remember that PD projects promote change that hypothetically leaves some better off while worsening the situation for others. Arguing out of self-interest is, therefore, not always a bad thing. Sometimes, such arguments can lead to new insight and new ways to approach the design.

Change can also affect current power relations. To empower workers means a distribution of power from others. Power relations, hierarchies, statuses and roles are embedded in healthcare practices [28]. Power is exercised between practitioners internally in a healthcare organization and between organizational boundaries, representing barriers to effective coordination and knowledge sharing [28]. Tension and conflicts may arise if the design implies changes to current balances. In Valkyrie, it is necessary to understand current power relations since we are working with different groups of healthcare professionals and with providers at different care levels. Likewise, it is crucial to understand how design changes might affect current power relations, hierarchies, statuses and roles. The theory of practice architecture can provide us with both a framework and a vocabulary to study the empirical relatings, sayings and doings at particular locations. Learning of these relations can move to broader areas to design a HIS to promote service coordination for a healthcare network. However, the theory of practice architecture will not solve this alone. After all, the willingness to design for change rests with the practitioners, which is why we want to involve them as active co-designers in Valkyrie.

*2) Selection of practitioners as co-designers*

It is crucial to involve suitable participants in the project since PD relies heavily on expert users as active co-designers. Ideally, participants should both represent their profession and be open-minded to learn from others to contribute to the design of their future work.

Stakeholder analysis, a method used in PD [13], supports the discovery of relevant stakeholders influencing or influenced by, directly or indirectly, the design project. Stakeholder analysis is relevant to Valkyrie, but finding relevant stakeholders in a healthcare network gets complicated since we also need to consider the organizational boundaries. We believe that the theory of practice architecture can contribute to the stakeholder analysis in at least two ways.

First, the focus on practices adds a layer to the stakeholder analysis, otherwise oriented towards individuals and organizational roles. This move introduces a domain-neutral perspective across boundaries, relevant for discovering stakeholders, especially those indirectly influenced by design.

Second, the practice layer is also a change tool to plan for future practices. The view of future practices can lead to the discovery of future users and stakeholders considered relevant to the project.

While the theory of practice architecture contributes to the stakeholder analysis to discover relevant stakeholders to Valkyrie, it does not provide any mechanism to choose individuals suitable as active co-designers in the project.

*3) Establishing an environment for mutual learning*

Both PD and the theory of practice architecture focus on *change* and *understanding*. However, they have different starting points. The theory of practice architecture starts with understanding the mutual shaping and reshaping of practices and the practice architecture and eventually uses this insight for possible future rearrangement and change. PD has the change as a starting point, while understanding becomes a natural part of mutual learning during the design process.

To understand participants' work practices and needs, designers must first learn their language and how they think and act, indicating that the initial exploration of work in PD may take up much time. However, time is often a limiting factor for clinicians, who otherwise would use their time on treating patients. In Valkyrie, we believe that an ethnographic workplace study with open-ended interviews and observations of clinicians in their natural work setting can contribute to a better environment for mutual learning later on in the PD project. Clinicians can benefit from this by reducing additional time, creating a more focused mutual learning process, and lowering the risk of frustration and tension.

However, how detailed the ethnographic study should be is a question to be explored further. This is also ongoing discussions in the PD research community. Ethnography and design research seem to overlap, which is why we find the Sociomaterial-Design approach appealing. This practice-oriented approach bridges ethnography with design research to incorporate understanding and learning with action and change. However, we miss the sensitivity towards organizational constructs such as roles, statuses, hierarchies, and power relations. We find this highly relevant in HIS-design for a healthcare network, such as for Valkyrie. The theory of practice architecture can provide us with such perspective at a particular site and in broader areas. However, the theory of practice architecture does not provide any intersection between understanding and design that Sociomaterial-Design provides. In Valkyrie, we will further explore how practice-oriented approaches can contribute to PD projects in the healthcare domain.

## IV. Conclusion and Future Work

In this paper, we have discussed some challenges related to Participatory Design (PD) when designing an information system for a health care network. Further, we have discussed how practice-oriented approaches can help mitigate these challenges.

Practice-oriented approaches do not mitigate all aspects of the recognized challenges. However, they can contribute with a vocabulary and a way of thinking that can help inform the design through different phases of the research project. They can also help to identify actual empirical connections between practices and materiality to interpret and explain practice variation.

One of the practice-oriented accounts, the theory of practice architecture, provides a framework to understand the practices' actions through language, activities, power, and politics. Power and politics are relevant in healthcare and essential to move the design from local sites to broader healthcare networks.

This paper contributes to practice-oriented approaches in technology design by adding the dimension of power and politics. This paper also contributes to the ongoing discussions in the research community on how to evolve PD to meet the future needs of ubiquitous healthcare data in patient-centric healthcare.

In order to understand how practice-oriented approaches can contribute to mitigating the challenges of PD, we will explore this further as part of the Valkyrie project.

## REFERENCES

[1] World Health Organization (WHO), "Depression and other common mental disorders: global health estimates," World Health Organization (WHO), Geneva, 2017.

[2] G. P. Martin, E. Sutton, J. Willars, and M. Dixon-Woods, "Frameworks for change in healthcare organisations: a formative evaluation of the NHS change model," Health services management research, vol. 26, no. 2-3, pp. 65-75, 2013.

[3] A. A. Wright and I. T. Katz, "Beyond burnout—redesigning care to restore meaning and sanity for physicians," N Engl J Med, vol. 378, no. 4, pp. 309-311, 2018.

[4] M. Berg, "Patient care information systems and health care work: a sociotechnical approach," International Journal of Medical Informatics, vol. 55, no. 2, pp. 87-101, 1999/08/01/, 1999.

[5] J. G. Anderson, C. E. Aydin, and S. J. Jay, Evaluating Health Care Information Systems: Methods and Applications: London: Sage Publications, 1993.

[6] M. M. Yusof, A. Papazafeiropoulou, R. J. Paul, and L. K. Stergioulas, "Investigating evaluation frameworks for health information systems," International journal of medical informatics, vol. 77, no. 6, pp. 377-385, 2008.

[7] C. Granja, W. Janssen, and M. A. Johansen, "Factors determining the success and failure of eHealth interventions: systematic review of the literature," Journal of medical Internet research, vol. 20, no. 5, pp. e10235, 2018.

[8] L. G. Tornatzky and K. J. Klein, "Innovation characteristics and innovation adoption-implementation: A meta-analysis of findings," IEEE Transactions on engineering management, no. 1, pp. 28-45, 1982.

[9] J. Blomberg, M. Burrell, and G. Guest, "An ethnographic approach to design," The human-computer interaction handbook: fundamentals, evolving technologies and emerging applications, pp. 964–986: L. Erlbaum Associates Inc., 2002.

[10] P. Ehn, "Work-oriented design of computer artifacts," Arbetslivscentrum, 1988.

[11] C. Frauenberger, J. Good, G. Fitzpatrick, and O. S. Iversen, "In pursuit of rigour and accountability in participatory design," International journal of human-computer studies, vol. 74, pp. 93-106, 2015.

[12] C. Spinuzzi, "The Methodology of Participatory Design," Technical Communication, vol. 52, pp. 163-174, 05/01, 2005.

[13] B. Friedman and P. H. Kahn Jr, "Human values, ethics, and design," The human-computer interaction handbook, pp. 1267-1292: CRC press, 2007.

[14] S. Bødker and M. Kyng, "Participatory design that matters—Facing the big issues," ACM Transactions on Computer-Human Interaction (TOCHI), vol. 25, no. 1, pp. 1-31, 2018.

[15] D. Nicolini, Practice theory, work, and organization: An introduction: OUP Oxford, 2012.

[16] M. S. Feldman and W. J. Orlikowski, "Theorizing practice and practicing theory," Organization science, vol. 22, no. 5, pp. 1240-1253, 2011.

[17] S. Kemmis, R. McTaggart, and R. Nixon, The Action Research Planner: Doing Critical Participatory Action Research, 2014.

[18] "Valkyrie - distributed service-oriented architecture for coordinated healthcare services," [Retrieved: February, 2022]; https://ehealthresearch.no/en/projects/valkyrie-distributed-service-oriented-architecture-for-coordinated-healthcare-services.

[19] K. Mahon, S. Francisco, and S. Kemmis, Exploring education and professional practice: Springer, 2017.

[20] P. Bjørn and C. Østerlund, Sociomaterial-Design: Bounding technologies in practice: Springer, 2014.

[21] F. Kensing and J. Blomberg, "Participatory design: Issues and concerns," Computer supported cooperative work (CSCW), vol. 7, no. 3, pp. 167-185, 1998.

[22] D. J. Blomberg and H. Karasti, "Ethnography: Positioning ethnography within participatory design," Routledge international handbook of participatory design, pp. 106-136: Routledge, 2012.

[23] P. M. Leonardi, "Materiality, sociomateriality, and socio-technical systems: What do these terms mean? How are they different? Do we need them," Materiality and organizing: Social interaction in a technological world, vol. 25, pp. 10.1093, 2012.

[24] S. Kemmis and T. J. Smith, Enabling Praxis: Challenges for Education: Brill, 2008.

[25] T. R. Schatzki, The site of the social: A philosophical account of the constitution of social life and change: Penn State Press, 2002.

[26] D. Nicolini, "Knowing in practice. The case of telemedicine," OLKC (Organizational Learning, Knowledge and Capabilities), Warwick, March, pp. 20-22, 2006.

[27] B. Doolin and L. McLeod, "Sociomateriality and boundary objects in information systems development," European Journal of Information Systems, vol. 21, no. 5, pp. 570-586, 2012.

[28] G. Currie and O. Suhomlinova, "The impact of institutional forces upon knowledge sharing in the UK NHS: the triumph of professional power and the inconsistency of policy," Public Administration, vol. 84, no. 1, pp. 1-30, 2006.

[29] C. Edwards-Groves and P. Grootenboer, "Praxis and the theory of practice architectures: Resources for re-envisioning English education," Australian Journal of Language and Literacy, The, vol. 38, no. 3, pp. 150-161, 2015.

[30] A. Edwards, Being an expert professional practitioner: The relational turn in expertise: Springer science & business media, 2010.

# Assessing Methods to Model Patient-Centric Care Pathways across Multiple Healthcare Systems

Sonja Cassidy[1], Øivind Skeidsvoll Solvang[1], Ove K. Lintvedt[2], Terje Solvoll[2,3], Conceição Granja[2,3]

[1] Department of Architecture and Innovation
Helse Vest IKT AS
Bergen, Norway

[2] Norwegian Centre for E-health Research
University Hospital of North Norway
Tromsø, Norway

[3] Faculty of Nursing and Health Sciences
Nord University
Bodø, Norway

e-mail: sonja.cassidy@helse-vest-ikt.no[1], oivind.skeidsvoll.solvang@helse-vest-ikt.no[1], ove.lintvedt@ehealthresearch.no[2], terje.solvoll@ehealthresearch.no[2,3], conceicao.granja@ehealthresearch.no[2,3]

***Abstract*—Clinical pathways have been promoted to maximize care coordination. Over the last decade, they have developed into patient-centric pathways, including the patient's needs and expectations of future health care. Integrated patient-centric pathways facilitate and encourage the coordination between multiple levels of care. Electronic Health Records (EHRs) contain information that could support integrated, patient-centric pathways. This paper aims to identify the challenges of modelling patient-centric pathways spanning primary and specialist care and provide guidelines for meeting those challenges. The study results show the lack of a standard definition of an integrated, patient-centric pathway and the various methods used to describe them. Finally, the study provides recommendations for a new approach for modelling EHR-supported, patient-centric pathways across the care continuum.**

***Keywords-patient-centric care; pathway modelling; integrated care; electronic health records.***

## I. INTRODUCTION

A rise in life expectancy has increased the need for more complex health care as the number of patients with multiple chronic conditions increases [1]. More resources are needed to provide the same level of care, challenging the health care services available for other vulnerable groups, such as young adults struggling with mental illness. Simultaneously, innovative drug treatments, therapies, and technology allow patients to receive care in their own homes. Health literacy amongst the general population, developed through media, has increased patient expectations [2] and changed the relationship between the patient and the healthcare provider and the understanding of patient-centred care [3]. Clinical pathways have been promoted as a response to meet the patient's needs and expectations of future health care and maximize care coordination [4]. Extensive research on implementing clinical pathways has also improved quality of life and reduced morbidity [5].

The health care sector has been through several reforms to implement pathways for different illnesses. Various terms for clinical pathways have been used, such as "clinical guidelines", "patient pathways", "care pathways", or "integrated care pathways" [6][7]. A common understanding defines a *clinical pathway* as standardized steps to rapidly diagnose the patient's illness and initiate treatment [8]. Clinical pathways can have different aims, such as integrating multidisciplinary teams (horizontal integration) or integrating services across primary, secondary and tertiary care (vertical integration) [9]. Care pathways are often seen as a "care coordination" tool to optimize the sequence of healthcare interventions performed by multidisciplinary teams across disease groups and health settings [7]. Though the terms are used interchangeably, care integration is aimed "to facilitate and encourage the coordination between levels of care" [10]. The term integrated care pathways, understood as vertically integrated, will be used in this paper.

The use of pathways as a "clinician-directed, patient-focused response" to coordinate care [11] is challenged by the changing understanding of what patient-focused care, also referred to as "patient-centred", "personalized", or "patient-centric" care, is. Conventionally, the patient focus has been interpreted within a biomedical framework, where the patient's illness is defined by a set of signs and symptoms [12]. In the late 1960ies, the patient focus included "providing information to the patient" [13]. Over time, the understanding of patient-centric care developed to include taking into account the whole person, exploring the patients' experience and ideas about the problem as a resource to guide the interaction, sharing power and agreeing on managing their illness [14]. Recent research, such as by Rand, Dunn, Slade, Upadhyaya, and Sheehan [15], argues that patient experience should be viewed as evidence in healthcare decision-making. Together, the understanding of "patient-centric" and the "integrated care pathways" defines the concept of patient-centric, integrated care pathways for the context of this paper.

Whereas the patient needs and demands can be said to "force the integration of services", the supply side, including technology and information systems, may facilitate it [10]. Access to Electronic Health Records (EHRs) is considered an essential component [1] of supporting patient-centric, integrated care pathways [16]. Guided by a multilevel, socio-technical lens, healthcare delivery can be represented on a system level (macro), a meso level, concerned with regional and local health services and community factors, and a micro level, representing every-day practices and patient perspectives [17]. At a macro level, sharing information can allow for better planning of services [18] and, at a meso level, contribute to integrating professional teams and healthcare organizations. At a micro level, increased access to EHRs could support the empowerment of patients and improve the quality of care [10].

In general, many approaches, modelling methods and tools have been used to describe pathways in healthcare; several adapted from other industries [19]. Traditional process-modelling methods, such as Business Process Modelling Notation (BPMN) [20] and Event-driven Process Chain (EPC) [21], have been used to describe work processes at a macro or meso level. Lean Design principles have been used to model meso-level pathways in quality improvement projects [22]. Key Service Design concepts have been used for transformative healthcare service research, viewing knowledge of patients, their families and healthcare professionals as essential to understanding the healthcare service needs. The Service Design approach has also evolved to address technology-enabled services [23][24]. "Customer Journey Mapping" (CJM), referred to as "Patient Journey Mapping" (PJM), when applied in healthcare, represents a micro-level, patient-centric perspective of the health care service based on the patient narrative [25]. Customer Journey Mapping Language (CJML) is a more recently developed framework for customer journey analysis [26]. Data-driven pathway modelling, such as Process Mining, has been used to support building pathways by using existing data from EHRs to display the actual patient pathway [27]. Unified Modelling Language (UML) has provided a more system-centric modelling method, connecting stakeholders, tasks and system intended functionality [28].

While different methods are available for modelling each of the health service levels, to the best of our knowledge, there is no single modelling tool currently available that collectively addresses the macro, meso and micro levels. This paper provides a general overview of the research and methodological guidance on modelling integrated, patient-centric pathways across multiple levels of care by (1) identifying the challenges of modelling patient-centric pathways spanning primary and specialist care and (2) providing recommendations for dealing with these challenges.

The rest of this paper is organized as follows. Section II describes the literature search, and Section III describes different approaches to describe integrated patient-centric pathways. A discussion on how to combine methods to meet the challenges of modelling cross-sectional pathways is outlined in Section IV before concluding the paper in Section V.

## II. METHODS

A literature search was conducted using the PubMed and Scopus databases, restricted to the years 2006-2022. The databases were chosen to cover both the medical and technological sides of pathway modelling. The inclusion criteria for the search were studies describing how patient-centric pathways were modelled, published in English. This initial, broad search resulted in keywords used for a search in title, abstract and keywords narrowed down by using the terms "patient", "pathway" and "journey" combined with variations of the terms "centric" and "centred" to capture articles with a patient-centric approach when describing the pathways. To ensure the relevance of the articles to the problem described regarding care integration the terms "integrated", "coordinated", "collaborative", and "multidisciplinary" were added. Finally, variations of the terms "modelling", "developing", "mapping", constructing", "framework", or "method" were added to find literature that described pathway-modelling methods.

The PubMed search string was: ((patient-centered care[MeSH Major Topic]) OR ("patient-centric care")) AND ((Delivery of Health Care, Integrated) OR coordinated* OR collaborative OR multidisciplinary) AND (pathway* OR journey*) AND (model* OR framework or method*) AND (developing OR mapping or constructing). The Scopus search string TITLE-ABS-KEY(("patient-centered care" OR "patient-centric care") AND (integrated OR coordinated OR collaborative OR multidisciplinary) AND (pathway* OR journey*) AND (model* OR framework OR method*) AND (developing OR mapping OR constructing)) returned 31 articles, none which met the inclusion criteria. The terms related to integrated care were limiting the number of results and therefore removed. Based a search for relevant keywords to capture articles in Scopus regarding modelling methods two new terms were added: "pathway mapping" and "journey mapping". The final Scopus search string was: "pathway mapping" OR "journey mapping" AND patient AND pathway OR journey AND model* OR framework OR method* AND developing OR mapping OR constructing.

## III. RESULTS

The search returned a total of 127 journal articles. The lack of a common definition of an integrated, patient-centric pathway and standards for modelling them risks biasing the literature search results. The variation in terminology is characteristic of an emerging intersectional topic such as designing patient-centric care pathways spanning multiple healthcare levels, possibly supported by digital tools or/and EHR data. Conceptually, similar research can be reported in very different disciplines.

The articles were screened based on their abstracts. The inclusion criteria were: (1) methods for modelling were discussed (2) patients were engaged in developing the pathway models and (3) the pathways included multiple healthcare levels. The article had to include at least two of the criteria to be included. Clinical guidelines describing a workflow for a specific diagnosis in a hospital setting, though common in pathway research, were excluded as the research described in this paper focuses on integrated, patient-centric pathways across the care continuum.

The first criteria resulted in 88 articles being excluded. After full-text reviewing 39 articles, 28 were excluded, as they did not satisfy at least two of the inclusion criteria. In addition, two were excluded as they were by the same authors, describing the same methods, only used for different diseases. The remaining nine were included in this study [29][30][31][32][33][34][35][36][37].

Models on all of the macro, meso and micro levels presented earlier were represented in the research. Four studies discussed issues in regards to modelling pathways across multiple healthcare levels [32][34][36][37]. All the included studies described patients as being engaged in developing the pathways. Three studies used a participatory research design [30][35][37], actively involving patients throughout the development process, representing the shift to a more holistic patient-centric care. In these studies, data to build the pathways were obtained through observations, interviews, focus groups, surveys and workgroups, engaging patients and their care providers, and the clinicians' participation secured the meso-perspective in the pathway. Only one study used data from the EHR in a data-driven approach that contributed to understanding the information flow throughout the pathways [33]. The approach included using several IT artefacts, such as a smartphone app and home-based medical equipment, to provide data to an EHR for clinicians.

The patient-centric pathways, or patient journeys, were modelled using a variety of approaches, from general illustrations [29][34], spreadsheets [35] or diagrams [37], to some variation of a flowchart [29][32][34] or patient journey map [35]. One article described the use of software-development methodologies to design pathways [30]. Data gathered through review of medical charts was used to generate pathways in one article [36]. Two articles discussed methodological issues without presenting a model [31][33]. Process mining was discussed to complement pathway mapping, derivating patient pathways from electronic health records [33]. None of the articles included in the study presented the use of digital modelling tools develop the pathways.

Five studies mentioned digital technology and EHR data as critical components for successfully implementing patient-centric pathways [29][30][33][34][35]. Furthermore, the studies discussed how digital technology could be integrated into the patient pathway to support the exchange of health data in real-time, opening a proactive approach to care. One study developed system supported pathways through a secured web-based platform, including functions covering the end-to-end care process [30].

Three studies reported that existing patient pathway mapping tools required modifications to reflect patient journeys across multiple healthcare settings [30][35][37].

When developing the pathways, different objectives and perspectives on patient care also caused tension and obstacles between different levels of care, challenging the collaborative process of designing patient pathways [32]. Others described how a patient-centric approach to pathway modelling challenged knowledge silos, helped bridge disciplinary boundaries and provided a possibility to develop a common language around the multi-organizational pathway [35][37]. It was also argued that an EHR-supported, data-driven, or at least data-supported, patient-centric pathway could improve the continuity of care in a disaster situation such as the COVID-19 pandemic, as it would have permitted more precise management of the emergency response undertaken in primary and specialist care [34].

## IV. DISCUSSION

The first objective of this paper was to assess the challenges of modelling patient-centric pathways across multiple healthcare levels. Though the issue has been described theoretically in several studies, this paper's literature search neither provided methods to develop such pathways nor guidelines to do so. Consequently, self-designing a method is considered a suitable approach for the second objective of this paper; providing recommendations in the form of guidelines for how to deal with these challenges. The characteristics of patient-centric, integrated, EHR-supported care pathways, as described earlier, become requirements for the recommended modelling approach.

On the macro and meso levels, process modelling, when using a common language and easily understandable notation, has been documented to offer a greater understanding of different stakeholder perspectives across disciplines and organizational levels in healthcare [19]. Though no relevant research was found on using process modelling methods to model integrated, data-supported patient-centric pathways as defined in this paper, process modelling is considered beneficial to enhance communication and gather consensus for work processes at a macro and meso level and across multiple healthcare organizations.

At the micro level, the Service Design approach, the Lean method and PJM all base the pathway models on patient narratives. Co-designing pathways with patients is a central element in these methods to empower patients. Including patients in the analysis and design of care pathways could also help identify gaps in the integrated care provision. However, PJM does not necessarily capture the clinical side of the patient pathway. Combining the patient perspective with principal activities in the clinical workflows as they are seen in the eyes of the healthcare professionals into one pathway model can aid in creating a shared understanding of the patient-centric pathway as a whole. Figure 1 presents a simple model using a combination of BPMN and CJML to develop a conceptual model to illustrate the complexity of modelling integrated, patient-centric care pathways. The model presents the start of care; a patient's initial meeting with their general practitioner (GP), the referral to specialized care, the health care levels involved, the healthcare professionals' activities, and the IT systems that supports these activities.

Digital modelling tools, such as MEGA, Qualiware X, BiZZdesign or Essential Project, are available for modelling pathways. Such tools allow sharing and reusing of models and can balance the trade-off between overview, comprehensiveness and detail. Recent expansions of BPMN, such as BPM+ Health, could add the possibility of deconstructing conceptual pathway models into more precise, possibly executable models, "connecting" the Patient

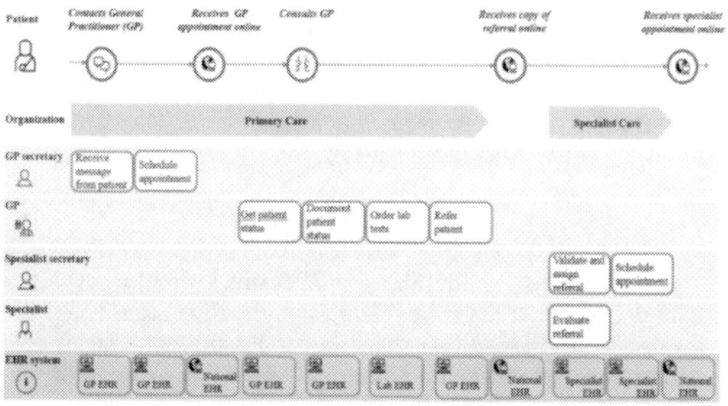


Figure 1. Example of a conceptual model of an patient-centric, EHR-supported pathway using BPMN and CJML. Adapted from [38].

Journey Map and BPMN process model to the EHR data [16]. This would represent the leap from a conceptual patient-centric pathway model to a digital one.

## V. CONCLUSION

The results from the literature search of this study present the variety of methods used to describe patient pathways and how the lack of a standardized methodological approach challenges the call for healthcare services transformation into more integrated, patient-centric care. We have suggested guidelines to meet those challenges, emphasizing the need to integrate the different stakeholder views and EHR data into one pathway model to provide an overview of essential elements in the pathway from a macro, meso and micro level perspective.

Combining BPMN, CJML, and BPM+ methods can help overcome the limitations of a specific method and provide a different healthcare research approach that meets the request from earlier research to design healthcare service experiences that qualify patient-centeredness, care integration and the use of EHR data to support the pathways. Further work is needed to develop method formalism, improve the visual presentation of the models' different perspectives, and exploit the potential of digital tools and reusable patient pathway templates. Finally, there is a need to explore how data from EHRs, regardless of where the EHR is located, can support integrated, patient-centric pathways.

## REFERENCES


[1] Organization for Economic Co-operation and Development (OECD) *Health at a Glance 2021: OECD Indicators*, pp. 94-95, OECD Publishing, Paris, 2021.

[2] World Health Organization (WHO) *WHO guideline: recommendations on digital interventions for health system strengthening: web supplement 2: summary of findings and GRADE tables*, WHO, Geneva, 2019.

[3] T. Smith, "Changing relationships between people and providers: making sense of patient centred health care," Quality & Safety in Health Care , vol. 15, no. 1, pp. 71-73, 2006, doi:10.1136/qshc.2005.017251.

[4] World Health Organization (WHO) *Framework on integrated, people-centred health services: report by the Secretariat*, WHO, Geneva, 2016.

[5] E. Aspland, D. Gartner, and P. Harper, "Clinical pathway modelling: a literature review," Health Systems (Basingstoke. England) , vol. 10, no. 1, pp. 1-23, 2019, doi:10.1080/20476965.2019.1652547.

[6] L. De Bleser et al., "Defining pathways," Journal of Nursing Management, vol. 14, no. 7, pp. 553-563, 2006.

[7] L. Bragato and K. Jacobs, "Care pathways: the road to better health services?," Journal of Health Organization and Management, vol. 17, no. 3, pp. 164-180, 2003, doi: 10.1108/14777260310480721.

[8] Norwegian Directorate of Health *National plan for implementing patient pathways for mental health and addictions 2018-2020*. Nasjonal plan for implementering av pakkeforløp for psykisk helse og rus 2018-2020 Helsedirektoratet, 2018.

[9] D. A. Conrad and W. L. Dowling, "Vertical integration in health services: theory and managerial implications," Health Care Management Review, vol. 15, no. 4, pp. 9-22, 1990, doi:10.1097/00004010-199001540-00003.

[10] O. Gröne and M. Garcia-Barbero, "Integrated care: a position paper of the WHO European Office for Integrated Health Care Services",International Journal of Integrated Care 1, vol. 1, 2001.

[11] S. D. Pearson, D. Goulart-Fisher, and T. H. Lee, "Critical pathways as a strategy for improving care: problems and potential," Annals of Internal Medicine, vol. 123, no. 12, pp. 941-948, 1995, doi:10.7326/0003-4819-123-12-199512150-00008.

[12] N. Mead and P. Bower, "Patient-centredness: a conceptual framework and review of the empirical literature," Social Science & Medicine, vol. 51, no. 7, pp. 1087-1110, 2000, doi:10.1016/s0277-9536(00)00098-8.

[13] H. Winefield, T. Murrell, J. Clifford, and E. Farmer, "The search for reliable and valid measures of patient-centredness," Psychology & Health, vol. 11, no. 6, pp. 811-824, 1996.

[14] M. Stewart, "Towards a global definition of patient centred care," British Medical Journal Publishing Group, vol. 322, 7284pp. 444-445, 2001, doi:10.1136/bmj.322.7284.444.

[15] L. Rand, M. Dunn, I. Slade, S. Upadhyaya, and M. Sheehan, "Understanding and using patient experiences as evidence in healthcare priority setting," Cost Effectiveness and Resource Allocation, vol. 17, no. 20, 2019, DOI: 10.1186/s12962-019-0188-1.

[16] T. Scheplitz, "Pathway Supporting Health Information Systems: Interdisciplinary Goal Integration - A Review," Innovation Through Information Systems, vol. 279, pp. 79-87, 2021, doi.org/10.3233/SHTI210093.

[17] T. Smith, K. McNeil, R. Mitchell, B. Boyle, and N. Ries, "A study of macro-, meso- and micro-barriers and enablers affecting extended scopes of practice: the case of rural nurse practitioners in Australia," BMC Nursing, vol. 18, no. 1, pp. 14, 2019.

[18] S. Ferrante, S. Bonacina, G. Pozzi, F. Pinciroli, and S. Marceglia, "A Design Methodology for Medical Processes," Applied Clinical Informatics, vol. 7, no. 1, pp. 191-210, 2016, doi:10.4338/ACI-2015-08-RA-0111.

[19] G. Antonacci, L. Lennox, J. Barlow, L. Evans, and J. Reed, "Process mapping in healthcare: a systematic review," BMC Health Services Research, vol. 21, no. 1, pp. 1-15, 2021, doi:10.1186/s12913-021-06254-1.

[20] M. Burwitz, H. Schlieter, and W. Esswein, "Modeling clinical pathways-design and application of a domain-specific modeling language," Wirtschaftsinformatik Proceedings, pp. 1325-1339, 2013.

[21] C. Chiao, V. Künzle, and M. Reichert, "Integrated modeling of process- and data-centric software systems with PHILharmonicFlows," IEEE 1st International Workshop on Communicating Business Process and Software Models Quality, Understandability, and Maintainability (CPSM), pp. 1-10, 2013, doi:10.1109/CPSM.2013.6703085.

[22] M. Malmbrandt and P. Åhlström, "An instrument for assessing lean service adoption," International Journal of Operations & Production Management, vol. 33, no. 9, 2013, doi:10.1108/IJOPM-05-2011-0175.

[23] J. Teixeira, L. Patrício, and T. Tuunanen, "Advancing service design research with design science research," Journal of Service Management, vol. 30, no. 5, pp. 577-592, 2019, doi: 10.1108/JOSM-05-2019-0131.

[24] L. Patricio et al., "Leveraging service design for healthcare transformation: toward people-centered, integrated, and technology-enabled healthcare systems," Journal of Service Management, vol. 31, no. 5, pp. 889-909, 2020, doi:10.1108/JOSM-11-2019-0332.

[25] A. Følstad and K. Kvale, "Customer journeys: a systematic literature review," Journal of Service Theory and Practice, vol. 28, issue 2, pp. 196-227, 2018, doi:10.1108/JSTP-11-2014-0261.

[26] R. Halvorsrud, K. Kvale, and A. Følstad, "Improving service quality through customer journey analysis," Journal of Service Theory and Practice, vol. 26, issue 6, pp. 840-867, 2016, doi:10.1108/JSTP-05-2015-0111.

[27] A. W. Kempa-Liehr et al., "Healthcare pathway discovery and probabilistic machine learning," International Journal of Medical Informatics, vol. 137, 104087, 2020, doi: 10.1016/j.ijmedinf.2020.104087.

[28] M. Askari et al., "A combined disease management and process modeling approach for assessing and improving care processes: a fall management case-study," International journal of medical informatics, vol. 82, no. 10, pp. 1022-33, Oct 2013, doi: 10.1016/j.ijmedinf.2013.06.011.

[29] M. Hägglund, P. Bolin, and S. Koch, "Experiences as input to eHealth design - a hip surgery patient journey case," Studies in health technology and informatics, vol. 210, pp. 672-4, 2015.

[30] N. Dubuc et al., "Computerized Care-Pathways (CCPs) System to Support Person-Centered, Integrated, and Proactive Care in Home-Care Settings," Informatics for Health & Social Care, vol. 46, no. 1, pp. 100-111, Mar 2, 2021, doi: 10.1080/17538157.2020.1865969.

[31] S. C. Madathil, A. J. Lopes, and M. Alfred, "Patient Journey Mapping: A Literature Review," IIE Annual Conference. Proceedings, Norcross, 2020, pp. 937-942.

[32] T. Røsstad, H. Garåsen, A. Steinsbekk, O. Sletvold, and A. Grimsmo, "Development of a patient-centred care pathway across healthcare providers: a qualitative study," BMC Health Services Research, vol. 13, 121, 2013, doi: 10.1186/1472-6963-13-121.

[33] M. Manktelow, A. Iftikhar, M. Bucholc, M. McCann, and M. O'Kane, "Clinical and operational insights from data-driven care pathway mapping: a systematic review," BMC Medical Informatics and Decision Making, vol. 22, no. 1, 2022, doi: 10.1186/s12911-022-01756-2.

[34] R. Devi, K. Kanitkar, R. Narendhar, K. Sehmi, and K. Subramaniam, "A Narrative Review of the Patient Journey Through the Lens of Non-communicable Diseases in Low- and Middle-Income Countries," Advances in therapy, vol. 37, no. 12, pp. 4808-4830, Dec 2020, doi: 10.1007/s12325-020-01519-3.

[35] S. McCarthy et al., "Embedding the Pillars of Quality in Health Information Technology Solutions Using "Integrated Patient Journey Mapping" (IPJM): Case Study," JMIR Human Factors, vol. 7, no. 3, pp. e17416, Sep 17, 2020, doi:10.2196/17416.

[36] N. Flora, H. Barbaree, A. I. Simpson, S. Noh, and K. McKenzie, "Pathways to forensic mental health care in Toronto: a comparison of European, African-Caribbean, and other ethnoracial groups in Toronto," Canadian journal of psychiatry. Revue canadienne de psychiatrie, vol. 57, no. 7, pp. 414-21, Jul 2012, doi: 10.1177/070674371205700704

[37] J. Kelly, J. Dwyer, T. Mackean, O. Donnell K, and E. Willis, "Coproducing Aboriginal patient journey mapping tools for improved quality and coordination of care," Australian Journal of Primary Health, vol. 23, no. 6, pp. 536-542, Dec, 2017, doi: 10.1071/PY16069

[38] R. Halvorsrud, I. Haugstveit, A. Pultier. "Evaluation of a modelling language for customer journeys" IEEE Symposium on Visual Languages and Human-Centric Computing (VL/HCC), Cambridge, UK, 5-7. Sept 2016 pp. 40-48, 2016.